Agricultural
Safety

Agricultural Safety

Keith E. Barenklau

CRC Press
Taylor & Francis Group
Boca Raton London New York

CRC Press is an imprint of the
Taylor & Francis Group, an **informa** business

CRC Press
Taylor & Francis Group
6000 Broken Sound Parkway NW, Suite 300
Boca Raton, FL 33487-2742

First issued in paperback 2019

© 1993 by Taylor & Francis Group, LLC
CRC Press is an imprint of Taylor & Francis Group, an Informa business

No claim to original U.S. Government works

ISBN-13: 978-1-56670-487-8 (hbk)
ISBN-13: 978-0-367-39733-3 (pbk)
International Standard Book Number 1-56670-487-1

This book contains information obtained from authentic and highly regarded sources. Reasonable efforts have been made to publish reliable data and information, but the author and publisher cannot assume responsibility for the validity of all materials or the consequences of their use. The authors and publishers have attempted to trace the copyright holders of all material reproduced in this publication and apologize to copyright holders if permission to publish in this form has not been obtained. If any copyright material has not been acknowledged please write and let us know so we may rectify in any future reprint.

Library of Congress Cataloging-in-Publication Data

Barenklau, Keith.
    Agricultural safety / Keith Barenklau.
        p.   cm.
    Includes bibliographical references (p.).
    ISBN 1-56670-487-1 (alk. paper)
    1. Agriculture--Safety. I. Title.

    S565 .B37 2001
    630'.2'089—dc21   2001018466

Visit the Taylor & Francis Web site at
http://www.taylorandfrancis.com

and the CRC Press Web site at
http://www.crcpress.com

Dedication

To Dorothy

Preface

Somewhere between the size of the small family farms of the 1930s and 1940s and the dozen or more large, employee-operated, specialized farms of the past couple of decades lies the farm for which this book is written. The farm envisioned in this book is family owned. The father, the proprietor, operates this farm with his wife, a married son who lives in town, a full-time employee, and up to three casual laborers, depending upon the season of the year. The farm has 1800 acres under cultivation, with some 300 acres of pastureland. The farmstead plants and harvests corn, oats, soy beans, and sometimes flax. The farm markets about 350 fat cattle and 200 hogs per year. About one third of the cattle and all of the hogs are raised on-site. The farm has a small herd of sheep and raises a small number of chickens that are mainly sold as fryers. Much of the same type of machinery found on farms of this sort is found on this one. The farm also has a corn drying operation that is used for the farm's crops only. The farmed land is not contiguous. Most of the land is within 2 miles, but one 240-acre plot is about 5 miles from the house and buildings. Heavy equipment is occasionally moved by road. This farm is considered typical of those in operation today, perhaps found most often in the upper Midwest. Safety data for this type of operation might apply to many different kinds of farms, although dairy and range-cattle operations may experience hazards that might call for additional safety procedures besides those found in this book. Safety information contained herein should not be considered complete.

Even though data for losses due to accidents on farms are somewhat fragmented, it is obvious that agriculture is a dangerous business when compared with other industries within the country. The literature lists farming, mining, and construction as the three most hazardous trades. Mining is sometimes listed first in the danger category, a lead it sometimes relinquishes to farming. Data for losses due to accidents may be suspect from one situation to another, but farming uses the fewest methods and systems for collection of accident data. The lack of mandated collection systems has led some state agricultural colleges to compile accident data from reviews of regional newspapers. Why the lack of data collection systems for farms? This question has no ready answer. Compared with most industrial operations, farmworkers comprise a relatively small number of employees. Until recently, no research funds were available for database establishment in the farm safety area. The National Institute for Occupational Safety and Health (NIOSH) has provided some funding in recent years, and from the results of its funded studies a bit more accident loss information pertaining to farms has become available.

Since few safety programs/processes are available for farms, much of the data presented in this book came from lessons learned in other industries. Much of the data from these industries is applicable to farms. There may even be an advantage in being a latecomer in agricultural safety: there is little to unlearn. Agriculture may benefit from the lessons learned over the years by the manufacturing and construction industries, without the fits and starts that the latter have had to go through.

The agricultural industry has no formalized accident databank from which it can obtain information. The Occupational Safety and Health Administration (OSHA) has no jurisdiction over farms with fewer than 11 workers, which effectively eliminates that agency as an information source. State agricultural colleges/schools have amassed a goodly amount of agricultural safety data, much of which may be found on the Internet. These schools are a good source for regional safety data, which is the type that should be of most use to farm proprietors. A grain and hog farmer in a midwestern state has perhaps little interest in what is happening in safety issues on the huge southwestern cattle ranches and vice versa. The individual farm proprietor has little need to lament the lack of an accurate national farm safety database.

In farming, as in other industries, the boss is the key person responsible for safety. In many industries the boss is called a manager, CEO, COO, etc. In this book, the top manager of the farm is called the proprietor.

No single book, including this one, can address all the safety concerns of a farm enterprise. An effort has been made, however, to present some useful information to help those wishing to achieve a more accident-free operation.

This book is presented in two parts. Part I includes some of the accident prevention concerns that may apply to most farming operations, along with actions that may be taken to help reduce accidental losses. Part II covers some of the safety work activities designed to help proprietors become more proactive in their quest to prevent accidents.

Table of Contents

Chapter 4
Fire Safety...37

Chapter 5
Machine Safety..43

Chapter 6
Environmental Safety on the Farm 57

Part II
Activities Designed to Assist the Farm Proprietor in Preventing
Accidents by Promoting Work Safety

Further Reading Suggestions

Suggestions for further readings are provided at the end of each chapter. Unlike industrial safety where often dozens of items are available, agricultural safety has far fewer sources. References come from three major sources in agricultural safety:

- Published books
- Magazine articles
- The Internet

Of these sources, the Internet provides the greatest number as well as the most current. By using the keywords agricultural safety, or farm safety, the searcher will find many references. State departments of agriculture appear to use the Internet a lot to share information. No attempt has been made to provide a long, comprehensive listing of Internet entries on various safety subjects. Those who use the Internet for further research will not be disappointed by the quantity of safety materials available. Books and magazine articles provide good information, but rarely is this information written by and for agricultural safety people. Information found will often be considered industrial safety data and will be modified to apply to the agricultural scene.

A review of the literature reveals few books in print on agricultural safety. A major reason for the scarcity may be that few governmental mandates are found in the farm safety area.

Like most books, magazine articles also slant the safety message in industrial terms. Often the question needs to be asked, "How can this safety data be adapted for agricultural use?" Much safety data is directly transferable from the factory or transportation company. For example, electrical current overloads provide injury or damage potential on the livestock feeding floor as well as in the welding bay of a metal fabrication shop.

PART I

Accident Concerns and Actions That May Be Taken to Reduce Accident Losses in the Farming Community

CHAPTER 1

Introduction to Agricultural Safety

The purpose of this book is to assist agricultural proprietors and workers in their efforts to avoid accidental loss. Many types of accidental losses can occur.

ACCIDENTAL LOSSES

- Physical harm to people and farm animals
- Damage to machinery, buildings, and other equipment
- Degradation of the products being produced
- Untimely interruption of processes
- Degradation of the local and gross environment

Many people think of accidents and injuries as synonymous. These thoughts limit their ability to see the larger picture and may prevent them from taking preventive action on other types of losses. Injury is only one possible result of an accident. The term physical harm, used above, is a more inclusive term which relates to both traumatic injury and disease as well as to other physical, neurological, and systemic effects which may be brought about through contact with elements within the environment.

Generally, damage accidents are more numerous than injury accidents and usually involve more expense. Industrial reports have shown that damage costs may run as high as 50:1 as compared with injury costs in capital-intensive businesses. This rather wide cost spread may not be completely applicable for farms, but it would be well to bear in mind the high cost of damage accidents. Farm proprietors own the capital equipment and are therefore in a better position to appreciate the costs of damage than an employee who works for a salary and has no ownership of the tools and processes.

Accidents may degrade the products produced as well as the services provided. Few data exist on cost comparisons for this type of loss, but the losses are very real in agriculture. An example of an accident might be that the wind blew off an unsecured grain bin cover and a rainstorm soaked several hundred bushels of stored grain. The loss, in dollars, could be considerable: for example, bad grain spoils good grain. The bin may have to be emptied, the wet grain dried, and the bin refilled. The dollar loss here is in extra time spent.

Process interruption caused by an accident may be a costly affair. For example: a truck that was needed to deliver poultry to a processor by 6:00 p.m. today was involved in a small traffic accident while in town on another job. The taillights were damaged, making the truck unsafe to operate at night. It was after 6:00 p.m. when the lights were finally repaired, but before the chickens could be loaded the processing plant called and cancelled its order. It purchased the chickens elsewhere.

The loss in a situation like this is hard to estimate. Perhaps another processor will take the chickens later; meanwhile, they will have to be fed and cared for. The price they fetch may be less on the open market than on the signed order. The proprietor's reputation may suffer — particularly with the processing plant.

Live agricultural products are hard to store. When they are ready for market they need to be sold. Keeping them around costs money. Overweight butcher hogs sell at a discounted price, in spite of the fact that the proprietor has additional feed costs to bear. Often, a small accident may cause a chain of events to take place that add up to a much greater loss than the original mishap. These are process interruption losses.

Accidents that degrade the environment can be very costly. Because of environmental concerns and laws, proprietors no longer have the freedom to dispose of waste products that they once had. It is probably true that relatively few farm proprietors would knowingly pollute the environment. Accidental pollution, however, can happen. For example, two barrels of used oil fall off the flatbed truck on the way to the oil disposal facility in town. They fall into a ditch with running water in it, and rupture. The oil flowed with the water into a neighbor's stock pond. This is a liability-type loss, as are many environmental losses. Working with the neighbor, hiring a waste cleanup firm, and making arrangements for the neighbor's cattle to be watered elsewhere can be an expensive proposition. The local, state, or federal environmental or transportation authorities may get involved. There may be citations and fines to pay. And the time spent on this type of accident is a cost to the farm proprietor.

LIABILITY-TYPE LOSSES

Liability-type losses are far more common today than they were a generation ago. Lawsuits are more common now than before. This type of loss is often a very costly one. Awards made by courts have increased dramatically. It is perhaps correct to say that winning a liability suit today costs more than losing one a decade or so ago. Legal fees, expert witness, and other trial expenses add up very quickly. The dollar sizes of these losses are hard to predict, since it is usually a judge and a jury who determine the amount. Proprietors can insure their operations to protect themselves from liability losses, but the cost of insurance is also a factor that affects the profitability of an agricultural operation. Liability losses almost always involve a third party. Following are some liability losses that have taken place involving farm operations.

- Farm equipment left in a position where people may get hurt in connection with it has been classified as an attractive nuisance in some court cases. Equipment left in a field beside a road may attract children and others to climb around on it and perhaps get hurt. Parents have been known to successfully sue the proprietor for leaving equipment in such an inviting place.
- Unmarked and/or unguarded holes in the ground that are a part of a construction project may result in liability loss if a third party falls into one. While it may be true that the proprietor did not invite these people to visit his farm, he still may be held liable. In fact, some people who have come onto the farm with the intent of committing mischief or a crime have been injured, and the court found in favor of the injured parties.

Perhaps things of this sort have caused proprietors to consider leaving the farm and engaging in some other enterprise! Even mailboxes can be a source of liability loss. Passing vehicles occasionally collide with roadside mailboxes. If the mailbox is installed too close to the roadway, a liability action may ensue. Letter carriers can usually provide regulations for proper installation of these receptacles. Proprietors would do well to consider the effects of all types of liability potentials when doing their safety planning.

Why this rather lengthy discussion of losses? The data will assist farm people in broadening their definition of accidental losses. The way people define things in large part determines how they are going to deal with them. If proprietors look upon accidents as injury events only, they tend to limit their responses to situations. These limitations can be very costly. Prevention of all types of losses should be a primary consideration.

FARM ACCIDENT STATISTICS

Some injury statistics are available for agricultural operations, but data on damage and liability losses are very hard to come by. It is known that the accidental death rate in agriculture enterprises ranks closely with other dangerous operations, such as mining and construction. Agriculture, mining, and construction have the highest accidental death rates of the major industries. Depending upon which set of statistics is consulted, sometimes agriculture shows up as the industry with the highest death rate, followed by mining. In other statistics, the reverse appears to be true. One statistic for 1996 shows 21 accident-related deaths per 100,000 farmworkers. During the same year, mining had 25 deaths per 100,000. Thus, in 1996, the accidental death rate in the mining industry was greater than that of agriculture. Other statistics for recent years show agriculture to have a higher rate than mining. In any case, when all industries are combined, in 1996 there were 4 deaths per 100,000 workers. It could be said, on the basis of accidental deaths and the information presented above, that agriculture is at least five times more dangerous than the average industry.

Another statistic that points out the danger of agricultural work is that farmworkers comprise only about 3% of the American workforce, but account for some 10% of the fatal injuries. These national statistics, which could be suspect to some degree, are perhaps meaningful to proprietors. The fact that mining has steadily shown improvement in this regard over the years, and agriculture has not, is perhaps more meaningful.

Why an improvement in mining and not in farming? There may be no fully suitable answer to this. Some indicators, however, may be considered. The dangers connected with mining have been a national concern in this country for a number of years. The Mine Safety and Health Administration (MSHA), a federal agency, has studied mine safety concerns and has engaged in many programs and initiatives designed to make mining safer. Fines are levied against mines that do not measure up to certain standards of safety. A lot of effort has been spent in safety education for miners. A mine safety and health academy in West Virginia is staffed by industry experts who have worked to assist in making mines safer places in which to work. Agriculture has not received anywhere near this level of support. The existence of OSHA has prompted factory owners and management to reevaluate their safety efforts. Agriculture, though it has a cabinet department, has not had a government enforcement agency looking over its shoulders as far as safety is concerned. The old squeaky wheel axiom has perhaps been at work here.

The intention is not to imply that agriculture has nothing going for it as far as safety is concerned. Within the past few years, the National Institute of Occupational Safety and Health (NIOSH) has provided some funding to study the agricultural accident problem, and both private and governmental regional groups have made some inroads. Local agricultural departments at colleges and universities have done noteworthy research on behalf of the farm population. These departments provide the most easy-to-reach help available today for the farm proprietor who is working to build a safer operation. A sizable amount of agricultural safety information may be found on the Internet, much of it provided by colleges and universities.

NONFATAL FARM ACCIDENTS

Various estimates place nonfatal farm injuries at one injury per five farms per year. These are injuries serious enough to require medical treatment and a time loss of at least a day. Again, because farms have no central reporting or recording agency, national information in the nonfatal injury area is sketchy. One statistic shows that in a recent year an estimated 200,000 work-related, nonfatal injuries occurred on farms. Some 65% of these injuries were suffered by owner/operator/partner personnel and the remaining 35% by hired farmworkers. Unlike general industry, usually the proprietors of farms are also primary laborers.

FACTORS AFFECTING THE FARM ACCIDENT PICTURE

The death and injury data paint a gloomy picture of safety on the farm. Why is farming so dangerous? Many things can influence statistics. One possible reason for the higher injury and death rates may be that both the very young and very old reside on farms and are often subject to hazards found on site. Farm sites are not only workplaces, they are also living places. If the very young and very old lived in factories, their accident rates would perhaps be much higher. Children work on farms; they normally do not work in factories. Older people retire and continue to live on farms. This is not the case with factories. Nothing here is intended to suggest that proprietors are engaging in illegal child labor, or taking advantage of old people. The old and the young are simply there — within the operation and subject to its hazards. A 1987 NIOSH study found that more than 12,873 accidental injuries happened to farmworkers who were between the ages of 10 and 19. Of those injured, 89% were males.

FARMWORKERS ARE ON THEIR OWN

Much of the time, farmworkers work alone. Factory workers, on the other hand, usually have fellow employees nearby to assist them if they need it, and usually a foreman is within shouting distance if a worker gets into trouble. Normally, the farmworker has only his/her own resources to call upon in a dangerous situation. These circumstances call for pre-job planning. Even a few minutes of discussion between the proprietor and worker, reviewing the known hazards of the job, would be worthwhile. Proprietors have little control over the work environment. They cannot control the weather, conditions of light and darkness, ambient temperatures, wind, dust, and the like.

Farmworkers are required to do a variety of chores. Many like farm work for this reason. A worker may be welding first thing in the morning, driving a truck later in the forenoon, working with animals before lunch, digging some post holes after that, and perhaps clearing some fallen tree limbs with a chain saw before supper. After supper he/she may assist with some animal vaccinations before retiring for the day. Farmworkers do a lot of things in a workday that might make the average factory worker uncomfortable. Different hazards are connected with each task a farmworker performs. Workdays may last until after dark, or longer. All the jobs, hours, and environmental conditions impact upon the farmworker. Specialized labor does not exist on the farm to the extent it does in many other lines of work.

MANAGEMENT OF RISK

To say that farming is a risky business is perhaps an understatement. A proprietor faces many types of risk during a given year. Risks such as the amount of rainfall, the threat of lower prices, whether a new brand of seed will produce as expected, and the like are very real, but will not be discussed here. This section will deal with the risks associated with accidents and accidental loss.

Many business entities have a person on the payroll with the title of Risk Manager. In addition to his/her many other duties, the farm proprietor must also function as a risk manager. What is risk management?

Risk management — *The work of preventing or reducing to the least possible cost losses which could strike an organization.*

What do risk managers do? They make decisions in line with certain risk management functions. They may choose to:

- Assume the risk — Even though a given job or part of the work is very dangerous, it may be deemed necessary.
- Avoid the risk — A certain task may have too high a risk for the benefits received, so it will no longer be done. Perhaps this risk can be transferred.
- Transfer the risk — This involves placing the risk with another entity or party, such as an insurance company, or hiring a specialized organization to do high risk work.
- Control the risk — This is where accident prevention activities come into play.

Obviously, professional risk managers have many duties and have formally studied that line of work. A lot more is involved in risk management than has been listed here. Farm proprietors, however, may employ some of the concepts and techniques used by those who make their living in risk management.

GREAT CARES OF FARM MANAGEMENT

Production decisions usually involve some degree of trade-off. Safety considerations are no different. A proprietor has many cares that concern him/her. It is an important duty of management to keep these cares in balance. What are the great cares of farm management? They are similar to those of almost any business. They may be summarized as:

- Production control
- Quality control
- Loss control (safety)
- Cost control

These cares should be considered in the decisions concerning the ongoing management of the farm. These cares are so closely interconnected that it is almost impossible to make a change in one of them without having some effect on the other three. For example, overemphasis on production will usually cause problems in quality and safety. Overemphasis on quality will usually cause production and cost problems. When safety is overly emphasized, production and costs suffer. Too heavy an emphasis on cost control has a dampening effect on the others. All four cares are important and they vie for the proprietor's attention. One of them is no more important than another, although one may be emphasized more strongly than the others in given situations. Consider an old saying in safety, "Safety First." Is safety first? It may be for a few minutes, a few hours, or even a few days in a major project, but it is not always first. Productivity and quality are normally first. Usually, if production, quality, and safety are under control, costs will be within desired bounds.

ACCIDENT COSTS

Accident costs are very real losses, and these losses are hard to make up. The cost of an accident immediately takes dollars from the credit side of the ledger to the debit side. These costs may necessitate postponement of the acquisition of tools, equipment, and other production needs. Making up the costs of a loss requires a lot of extra work and the need to produce more. Following is an example of how hard it is to make up for the costs of losses.

Cost of Loss	Profit Percentage	Sales Required to Make Up Loss
$5,000	10%	$50,000
$5,000	5%	$100,000
$5,000	2%	$250,000

The numbers speak for themselves. People connected to a farming operation know how hard it is to increase sales by $100,000 — the amount of extra sales required to break even if a farm generating a 5% profit suffers a $5000 loss. In order to reduce or eliminate accident costs, a farming operation needs to reduce accidents. Of course, not every accident costs $5000. Some cost less, but others cost more.

To this point, nothing has been said about the cost of human and animal suffering. It is difficult to place a dollar amount on this, but those who suffer the pain of injury pay a high price indeed. A great deal has been written about hidden costs. Perhaps no such thing exists. Some costs are hard to compute, but very few of these are hidden.

TYPES OF ACCIDENT COSTS

Insured ledger costs — These are costs which are covered by insurance. Proprietors pay a premium for insurance coverage that normally pays for an accidental loss, less deductibles, up to the limits specified in the insurance policy.

Uninsured ledger costs — These costs are for loss of assets which are on the farm's ledger, but are not covered by insurance. An example of this might be damage to a wagon and a corner post as a result of turning too sharply into a field. This cost may not be felt immediately, but will become apparent when the fence and wagon are fixed. Many machinery accidents fall into this category. Rarely do proprietors insure machinery, although some of it may be covered by fire insurance as an equipment rider attached to their general fire insurance policy. Some machine accident costs have to be paid soon after the accident, or additional costs may arise from having to lease a piece of equipment in order to finish a job.

Uninsured nonledger costs — Sometimes these are called hidden costs. They may involve having to drive around hunting for repair parts, or loss of time while the piece of equipment involved in the accident is being fixed. Time and transportation of injured workers for medical treatment are other uninsured nonledger costs. Making temporary repairs to buildings and grounds is another example.

Even costs paid by an insurance policy may come back to later present a problem. Many premiums are based upon the loss experience of the insured. Sometimes accidents cause premiums to increase. This is a rather common practice in auto insurance when one or more claims are filed in a rather short period of time. If losses become too high or too frequent, the insurance carrier may not only raise its rates, but may refuse to insure the operation at all. When this happens, it is often necessary to apply for coverage from a company that specializes in risks with higher losses. Premiums are usually much higher for this type of coverage. Keeping insurance costs down is another of the many reasons for loss prevention efforts on the farm.

SUMMARY

This chapter emphasizes that accident prevention is a necessity on today's farm. Most proprietors would never think of neglecting production, quality, or cost control efforts. Safety efforts are just as important. In fact, safety (accident control) is a full partner with the other three cares of agricultural management. Farm proprietors and workers have a few advantages that are not enjoyed by most factory workers. They are usually very well acquainted with the work they are doing and realize how it fits in with other things done for safe farm production. They are few in number, and for the most part they have worked together for a fairly long period of time. They perhaps have a closer personal relationship than many other workers do. It has been said that farmworkers seem to be able to anticipate what fellow workers are doing and why they do certain things in certain ways.

All those connected with a farm enterprise have to be aware of the potential for accidental losses, and know what to do in order to minimize their occurrence. The following chapters will provide information that will help farmworkers strengthen their safety efforts.

FURTHER READING

Anthony, S., *Farm and Ranch Safety Management,* Delmar Publishers, Albany, NY, 1995.
Bird, F. E., Jr., *Management Guide to Loss Control,* Institute Press, Loganville, GA, 1980.

CHAPTER 2

The Basics of Accident Prevention

The proprietor is responsible for everything the enterprise does, or fails to do.

— Anonymous

Certain basic principles are common to most types of safety programming. These basics have come about as a result of the studies done by safety professionals plus lessons learned by many proprietors and workers over the past several years. The key person in any entity's safety effort is the boss. Thus, on the farm, the proprietor is the key person. Safety begins with a boss who knows what he/she wants and is willing to provide the leadership required to get the job done. Most industrial organizations develop a safety policy statement that defines the accident problem, tells people what is to be done about it, and defines the authority/accountability process. The farm proprietor may do the same thing. This short document would set forth basic guidelines for all safety efforts in which the organization will engage. It may seem somewhat out of place to have a written safety policy statement for relatively few persons working on a given farm, but it is not. A statement of policy is necessary because it provides the basis for ongoing safety efforts.

SAFETY POLICY STATEMENT

A safety policy statement is a short (usually one page) document that is signed by the proprietor and given to all personnel. A copy of this statement is always given to a new worker as a part of his/her job orientation. The policy has four major parts:

- Definition of the accidental loss problem
- What is to be done about the problem
- Accountability for safe behavior
- Where to go for help

No matter how simple or complex the writing style, if the four parts shown above are included the policy will be a useful one. Following is an example of a farm safety policy statement:

Safety Policy — Donald's Farm

Our accident rate and costs are much too high. These costs are making profitability impossible and are eating into our cash reserves. If this trend continues we will not be able to continue operations, and that will mean a loss for everyone connected with our farm.

In order to correct this situation, all personnel concerned will have to work in a safe and competent manner, and follow the safety guidelines that we are developing for our use. We must operate more professionally than we have in the past and do the right things right.

I am responsible for safety on this farm. My son and partner will serve as safety director and will work with all of us to make sure that everyone is doing their part to turn the operation around for the better. All personnel are expected to follow the safety guidelines which have been and are being developed. Excuses for inactivity or inappropriate activity under the pretext of expediency, etc., will not be tolerated.

No one can make safety happen all alone. Therefore, everyone is asked to put their best foot forward as we move ahead to safer times. If you are not sure about something, ASK!

S/Bill Donald

The policy statement above is one that is in use today. Only the names were changed. Obviously, a one-page statement will not bring about safety by itself, but it serves as a statement of first principle as far as safety is concerned. The definition of safety as used in this book follows. It is viewed as synonymous with accident prevention.

Safety — *The work we do to conserve the resources of the organization and protect it from accidental loss.*

The word safety has a lot more meaning if it is thought of as a verb. Safety is something that is done in order to prevent accidents/losses. How does one do safety? Actually, safety is best done when it is accomplished just like production. A proprietor would never tell a hired hand to "go out there and raise cattle today." He would be a lot more specific than that. It is not uncommon, however, for the boss to say to a worker, "go out there and be safe today." How should he be safe? What should he do that he has not been doing? How can he be safer today than he was yesterday? If these questions are answered, then the boss has something meaningful to tell the worker about safety. Actually, safety is a lot like profit. No single work activity is known as profit. If there is, a lot of people are going to a lot of unnecessary work trying to farm! Profit is the result of something — usually getting more dollars out of an operation than was put into it. Getting the dollars out is the result of a lot of very specific work. Safety is like that. It is a result of doing the right things right and not getting into a set of circumstances where an accident is likely to happen. The following are ways to foster safety:

FOSTERING SAFETY

- *Decide what safety work is going to be done.* The work might be inspections, accident investigations, proper job training, giving safety suggestions, handling toxic substances safely, following certain rules when working with livestock, keeping guards in place on machinery, following proper electrical clearance procedures, and the like.
- *Establish guidelines for the work.* Safety work, like production work, is usually done as a series of steps. Guidelines help those who are going to do the work to plan and carry it out systematically. Making an inspection of machine guards, for example, is most efficiently done if a routine or checklist is followed. Guidelines provide assistance to those doing the work, and often will save time.
- *Check on progress.* The proprietors or designated safety representatives should oversee safety work closely enough so that they are sure the work is being done properly and in a timely manner. Progress checks will often reveal needed changes in the work guidelines.
- *Provide feedback to those doing the work.* Those checking on progress should always comment on exceptional work that meets or exceeds the guidelines, and on that which falls short of the mark. Comments reinforce the acceptable performances, and provide an opportunity to encourage those who are not following the guidelines.
- *Correct and revise guidelines as necessary.* Often, when guidelines are first used, rather frequent revisions are necessary. Later on they may be used without changes for some time. When a change is made in the safety work, the guideline should be revised to reflect that change. When all hands are doing a specific piece of safety work very well, that guideline may be dropped, and perhaps one added for an area where it is needed.

The construction and use of guidelines will be covered later in the book after some of the various safety work activity areas are explained. Use of guidelines is not new. Guidelines are often used in production activities and specialized operations like crop spraying. In safety, it is desirable to have everyone reading from the same script. Guidelines help achieve this.

FARM SAFETY RULES

Farm safety involves much more than adhering to a list of do's and don'ts. Safety rules should not be viewed as some sort of magic list which, if followed, leads to a utopian environment with little need for anything else. Rules, when properly constructed and followed, are time-savers that help to promote safety in the workplace. Rules may be grouped into two classes — general rules which apply to everyone on the farm, and specific rules which apply to certain tasks or jobs. General safety rules usually do not change much from year to year. Specific rules, on the other hand, may change often, depending upon the introduction of new machines, processes, and changes in the workplace.

Though it takes some time and effort to develop safety rules, they can save time in day-to-day operations. Rather than having to explain over and over again why certain things are done or not done, or done in certain ways, develop a rule that will cover the situation. Explain the rule to all that would fall under its jurisdiction. Questions about it need to be answered. Then, people should be asked to abide by the rule and corrected when they do not. Rules should be written down, posted where practical, and referred to prior to the start of major tasks. Rules may differ from one establishment to another, since hazards may be different at each. Following are examples of some safety rules. These may be modified to better fit a given operation.

General Safety Rules

1. Smoke only in designated and/or nonposted areas.
2. Alcoholic beverages and nonprescription drugs are not allowed on the job.
3. Report all on-the-job injuries to the proprietor or safety person.
4. Firearms may be carried only with the proprietor's permission.
5. Maintain order in your work area.
6. Report hazardous conditions promptly.
7. Follow safety precautions at all times.
8. Do not enter confined spaces without a standby person and permission.
9. Know where you can get help, if needed.
10. Horseplay is not allowed on the job.
11. Always bend your knees when you lift and get help with heavier loads.
12. Always wear personal protective equipment (PPE) on jobs requiring it.
13. Hats, long-sleeved shirts, and trousers are required for outdoor work.
14. Know where first-aid equipment is located.
15. Perform no emergency repairs unless authorized to do so.
16. Wear clothing appropriate for the temperature and weather conditions.
17. Secure all gates and doors properly.
18. Report unusual animal behavior.
19. Secure equipment and get inside during lightning and thunderstorms.
20. Be particularly alert during tornado season.

Specific Safety Rules

Electrical Safety

1. Make sure electrical power tools are grounded, if required.
2. Inspect power cords for safety and serviceability prior to use.
3. Attempt no electrical repairs unless you are qualified to do so.
4. Use the plug, not the cord, to unplug electrical equipment.
5. Know the location and proper use of electrical disconnects.
6. Keep equipment well away from overhead power lines.
7. Use nonconducting ladders when working near electrical energy sources.
8. Tag faulty electrical tools and place them in the repair basket.

9. Make sure of proper clearances when moving machinery under power lines.
10. If equipment, such as an elevator, has to be raised near power lines, get someone to serve as spotter.
11. Use no electrical power tools while standing in mud or water.
12. If unsure about electrical safety, ASK.

Fire Safety

1. Post "No Smoking" signs in fire-sensitive areas.
2. Keep all signs visible and in good repair.
3. Keep vegetation away from fire-sensitive areas such as fuel tanks and buildings.
4. Ensure that all flammable materials are labeled as such.
5. Keep flammable liquids in approved containers during transport or storage.
6. Do not allow combustible trash to accumulate.
7. Make sure fire fighting devices are fully charged and conveniently located.
8. Never fill vehicle tanks so that they run over.
9. Use care to keep combustible materials from getting on your clothing.
10. Stop engines before refueling.
11. Store oxygen and acetylene tanks separately and in a secured, upright position.
12. Keep safety caps in place on tanks until use.
13. Have fire protection available when welding or cutting.
14. If unsure about fire safety, ASK.

Tool Safety

1. Use the proper, serviceable tool for the job.
2. Use no cheater bars/pipes on wrenches.
3. Use the proper size of wrench/socket if available; adjustable wrenches are secondary tools.
4. Never remove safety guards/shields from power tools.
5. Do not use a power tool you are not familiar with, or one on which you have not been trained.
6. Rather than use unserviceable tools, place them in the repair basket and inform the proprietor of your action.
7. Use only serviceable ladders and inspect each before use.
8. Never lay a tool on the ground while in the field; it will be lost.
9. Keep power cords out of the path of vehicles.
10. Wear eye protection when grinding or sharpening.
11. If unsure about the use of a tool, ASK.

Machinery Safety

1. Operate no equipment unless you have been trained or had experience with it.
2. Allow no one to ride on equipment you are operating unless there is a passenger seat for him or her.
3. Make sure equipment is stopped before attempting to clean, service, or repair.
4. Keep guards in place.
5. Read and follow equipment manuals.
6. Always use the equipment manual or lubrication order when servicing equipment.

7. Lower all attachments before working on a machine.
8. Use safety blocks when suspending parts of machinery; using the jack alone is unsafe.
9. Perform a safety check before using equipment.
10. Use care when mounting and/or dismounting equipment.
11. Check onboard fire fighting devices, if so equipped.
12. Use extreme care when operating around ditches and embankments.
13. Make sure rollover protection structures (ROPS) are properly secured.
14. Check behind vehicles before backing.
15. Use the safety cage when changing tires on machinery.
16. Before you leave equipment, shut it down and lower all ground-engaging devices.
17. Do not wear loose or ragged clothing near rotating shafts or gear trains.
18. Modify equipment only with the manufacturer's approval.
19. Use extreme care when moving equipment on public roads.
20. Keep machinery off public roads at night, unless it is lighted in accordance with state law.
21. Be watchful and maintain full control when entering a public roadway or when entering or leaving a field through gates.
22. Make no sudden turns while operating in road gear, and stay on the traveled portion of the roadway.

Animal Safety

1. Remember that animals may be friendly, but they are extremely strong, and they spook much easier at night than in the daytime.
2. When with animals, stay out of corners and areas from which escape is difficult.
3. Bulls that are friendly can be very unpredictable when cows are in heat.
4. Be wary of newly acquired livestock.
5. Make sure all stock enclosures are in good repair.
6. Be sure to close gates securely.
7. Always report unusual animal behavior that could signal illness.
8. Request veterinary assistance for sick livestock.
9. Remove dead livestock as soon as possible.
10. Handle vaccines and medicines with care.
11. Work cattle with two people when possible.
12. If you are not sure about anything dealing with livestock, ASK.

Toxic Materials Safety

1. Never store animal vaccines or medicines in the family refrigerator.
2. Collect, store, and refer to the Material Safety Data Sheets (MSDS) which are shipped with most chemical compounds.
3. Make sure all employees who handle chemicals read the appropriate MSDS.
4. Read warning labels on containers and do not remove them.
5. Wear recommended PPE when handling or applying agricultural chemicals or insect sprays.
6. Be mindful of wind direction when preparing chemicals, so that people and animals are not exposed.
7. Inventory leftover chemicals yearly and dispose of them properly.

8. Notify mutual aid neighbors when you are about to use chemicals.
9. Avoid windy days when applying agricultural chemicals.
10. Dispose of all veterinary items properly.
11. If you have questions about toxic materials safety, ASK.

The above general and specific safety rules may be used as a guide when constructing rules for a given farming operation. Since accident exposures may differ from one farm to the next, rules should be developed with this in mind. Rules need to be updated from time to time. When a rule no longer applies, get rid of it. If exposures change, adding a new rule may be advisable. A point to remember: rules not only help to protect people when they know there is danger; they apply as well when people do not know a danger exists.

MUTUAL AID GROUPS

For years, many industrial plants have utilized the mutual aid concept as a part of their safety efforts. It may be an important consideration in farm neighborhoods. Mutual aid groups are safety oriented and a bit more formalized than the traditional neighbor helping neighbor practice. How do they work? Suppose there is a group of farms within a 5- to 6-mile area. Each of these farms perhaps has certain specialized tools, equipment, or know-how that, from time to time, would greatly assist another farm. Often, this equipment is expensive and used only once in awhile, but when needed it is very desirable to have. It would be economically impractical for each farm to invest in such equipment on its own. For example, farm A has a 1200-gallon water tank mounted on a road-type trailer. It is equipped with a high-capacity fire pump. The farms in the mutual aid group have a list of such equipment and maintain a list of what is available from each other. If farm B is planning to do a one- or two-day job that involves fire danger, it could request the use of the fire tank from farm A. Most groups have a rule that if the owner of the needed equipment is not using it, another group member may request it. Another mutually agreed-upon rule concerning the operation of the group, might be, the user who breaks it repairs it. A worker may have a skill that enables him/her to do something very well, such as finish-grading on a haul road. Skills as well as equipment may be listed. Someone acting as safety director for one of the farms usually maintains this list and it is distributed among the members.

Mutual aid groups have been used to further safety efforts for many years. They save money through the sharing of necessary, but seldom used, equipment. Rules governing a group may be jointly developed by representatives of the farms involved.

LEARNING FROM THE EXPERIENCE OF OTHERS

A person who lived on a farm for some 25 years recalled accidents that happened to his family, relatives, and neighbors during that time period. Valuable lessons can be learned through the suffering and loss of others. All the accidents and losses could have been prevented. It is too late for those involved, but others may learn from their experiences:

Big, friendly, and strong — A proprietor raised whiteface bulls on his farm. They were big, friendly animals. One day the proprietor was in the bullpen and was standing near a concrete silo wall. One of the bulls came up to him. Being playful, the bull nudged the man with his head. He forced the man against the concrete wall and fractured some ribs. The man fell to the ground, and the bull backed up and looked at him, as if to say, "What are you doing laying down there?" The bull was just being playful. His weight and power, however, caused some cracked ribs. Lesson: animals may be friendly, but they are very strong. Never get into a position where they can cause crush damage.

The horses were spooky that day — A proprietor suffered a broken hip and leg when horses attached to the manure spreader moved ahead when he was caught between the spreader wheel and the barn wall. Lesson: had the horses been tethered, or had the tugs been dropped, the accident would not have happened.

A piece flew off! — A proprietor was chiseling rivets from mower blade sections. A piece of metal broke off and struck him in the eye. He did not seek treatment, and eventually lost sight in that eye. Lesson: always wear goggles when cutting metal. Promptly seek medical treatment for injuries.

Tractor fatality — The proprietor's eldest son was returning to the farmyard on a tractor. He had been a mile down the road working in a field. It was nearly dark. The tractor had no lights. The man missed the driveway (he may have been blinded by oncoming headlights), the tractor tipped over and crushed him to death. Lesson: do not travel on public roads without proper lights. Slow down for driveways when traveling in road gear.

Lost a forearm — A proprietor was trying to clear a jam on a corn picker's snapping rolls. He was alone in the field picking corn. He left the power takeoff running so the snapping rolls were moving. He used a corn stalk to try to poke the jam into the in-running nip point. The snapping rolls grabbed the stalk he was holding, and before he could release it, the rolls pulled his right arm into the machine. He lost his right forearm. Lesson: always stop machinery before attempting to clean, maintain, or adjust it.

How do you stop this thing? — Two 10-year-old boys saw an airplane make a forced landing in a cornfield near their farm. They walked through the field to it. The pilot had gone for help in another direction. The boys managed to get the plane's engine started. When it began to move, they jumped out. The plane taxied through the field and enmeshed itself in a fence. It was a total loss. Lesson: planes are expensive when the court makes you buy them. And you don't even get the plane! The two boys could have been badly injured.

I thought it was a rabbit — A proprietor was harvesting small grain with a binder. His four-year-old son, without the father's knowledge, went out to help his dad harvest. Somehow he got into the path of the cycle bar, which cut him at the elbow and dragged him several feet before his father heard him yelling. The father reported that he thought he had come into contact with a rabbit. The child suffered severe cuts at his right elbow. Lesson: farm people, do you know where your children are? This injury could have been much worse.

Tractor ran over child — A child was killed when a tractor ran over him. He and his brother were playing on the parked tractor. One child was in front of one of the drive wheels when his brother released the brakes. The tractor rolled forward and killed the boy. Lesson: children should not be permitted to play on heavy equipment.

I thought the can had motor oil in it — A farmworker was told to put a quart of oil in a tractor before taking it to the field. He poured a quart of what he thought was oil from a 5-gallon can. It turned out to be weed spray concentrate. The proprietor noted that the oilcan was still full, so he went looking for the worker who had used the tractor for about 30 minutes with weed spray in the crankcase. He stopped the tractor, loaded it on a trailer, and took it to the neighborhood repair facility. A mechanic drained the crankcase, put a couple of gallons of soapy water in the engine, and ran the engine at a fast idle for several minutes. He then drained the crankcase again, flushed it with kerosene, and refilled it with the proper amount of engine oil. The tractor ran fine with no signs of engine damage. Lesson: do not use chemical containers for any purpose than that for which they were designed. Make sure all weed spray, insecticides, and the like have their labels clearly displayed. This incident could have resulted in a ruined engine.

Strongest wind we've seen — An empty rubber-tired wagon was blown by the wind on a gusty night. It rolled across the equipment yard and crashed into a grain elevator, badly damaging both pieces of equipment. Lesson: in areas subject to high winds, always chock wheels of empty wagons.

The above accidents, typical or not, happened in one farm community. Three of them involved no injury at all, but some rather expensive property damage. Accidents like these may be typical of many farm communities. Have the lessons learned been passed on to today's workers? It is less costly as well as less painful to learn from the mistakes of others. It is a good idea to review neighborhood accidents and discuss them with workers. This may prevent losses in the future.

Collection and Distribution of Accident Data

Often, mutual aid groups share accident data among members. When something happens to one of them, they write it up and distribute it to others. By such sharing, all may benefit from a single occurrence that happened to one of the members. Since the purpose is to assist others in accident prevention, the names can be changed or omitted. It serves no purpose to further embarrass involved persons.

NEAR MISS DATA

Some accidents do not involve injury or loss. They are sometimes called close calls or near misses. More of these events occur than those that actually cause loss. Why not record and share near miss data? In some safety literature, near misses are called incidents. Studies have shown that the same causes exist in incidents as they do in accidents. An incident can be reviewed in a calm manner since little or no emotion is involved. In an incident, nothing is damaged; nobody is hurt, but it is advantageous to study incidents. Sometimes incidents are laughed at. They are no laughing matter, however, because the next time the event occurs, there may be an injury or damage.

In order for incidents to be investigated and discussed, they have to be reported. They are easy to overlook, because there is no loss. Proprietors should ask workers to report near misses because of the wealth of accident prevention information that may be gleaned from them. Will workers report incidents? Most often they will — particularly if they are not punished for reporting the details. The proprietor need not mention the person's name. Incident investigation programs were used for many years in the U.S. Air Force. At one time their report form was called "AnyMouse." This word, of course, is an exact anagram for anonymous.

SUMMARY

Much of the material presented in this chapter is the result of lessons learned in heavy industry. Many industrial organizations have had safety programs and processes for a number of years. Much of their knowledge and findings are transferable to farm operations. Industry has an advantage because of its many years of dealing with loss prevention. The farming operations have an advantage in that the work forces are smaller and everyone usually knows everyone else. In addition, most workers on a given farm have worked together for several years and they are in a better position to appreciate how their job fits in with everyone else's. Because of the diverse nature of farm work, the laborers are perhaps exposed to a larger variety of hazards than their counterparts in industry. Industrial safety programs involve a goodly amount of paperwork. This need not be the case for safety programs on the farm. The small number and interdependence of farmworkers suggest that less paperwork may be needed.

However, an active pursuit of accident prevention on the farm is needed at this time. Though there may be few governmental agencies looking over the farmer's shoulder with respect to safety, statistics clearly indicate the need for better farm accident prevention. By working with their employees and perhaps a mutual aid group, farm proprietors can effect a lowering of accidental loss rates on their premises. A farm proprietor, near retirement, had something to add on the subject of farm safety, "We can't afford to take chances with safety in the future like we did in the past. The cost of failure today is too high."

FURTHER READING

Barenklau, K. E., Doing the Right Things Right, *Occupational Health & Safety,* Waco, TX, January, 1997.

Barenklau, K. E., Revitalizing Safety Management, *Occupational Health & Safety,* Waco, TX, March, 1999.

CHAPTER 3

Building and Grounds Safety

Hazards to the health and well-being of people and animals may be found in and around farm buildings and grounds. An additional responsibility is placed upon farm proprietors because the family and workers either live or spend many hours there. To the farm family, the buildings and grounds are a part of their home. In addition to growing crops and raising animals, the proprietor has responsibilities similar to those of a housing/property manager. Many farmworkers are adept at tasks normally thought of as belonging to the building and construction trades. Most have learned these farmstead maintenance skills while on the job.

Some of the farmstead tasks that the proprietor is responsible for include:

- Control of hazardous substances
- Repairs to buildings
- Carpentry, plumbing, and electrical work
- Control of vegetation and pests
- Constructing additions to buildings
- Confined space safety
- Haul road construction and maintenance
- Drainage
- Animal husbandry

The proprietor and his people may not be sufficiently skilled, or may not be able to do all these things in a timely manner; contractors are sometimes employed to do them. Many proprietors act as their own general contractor for work done on the farmstead. When using contractors, make sure the hired firm has sufficient liability insurance and workers' compensation insurance (if required) for their employees. The proprietor and his people should do all they can to warn or protect the contractor's workers from harm while on the farm.

Keep in mind that when a contractor's workers are injured on a proprietor's premises, liability-type losses are possible. Ask the contractor for copies of his certificates of insurance if in doubt as to his coverages. Be prepared to supply yours to the contractor, if requested. The proprietor's insurance carrier or agent can supply documents which show the farm's insurance coverages.

CONTROL OF HAZARDOUS SUBSTANCES

The protection of people and animals from hazardous chemicals and substances is an important safety consideration. Generally, farms with more than 11 employees fall under the jurisdiction of the Occupational Safety and Health Act (OSHA). This book is primarily addressed to farms with less than 11 employees who are not subject to OSHA laws. Even though a farm may not be subject to OSHA regulations, these regulations may be used as guides in the construction of many safety topics. One of these topics is hazardous substances. The OSHA regulation CFR 1910.1200 is commonly called the Hazardous Materials Communications Act. This regulation will be used as a source for this section dealing with protecting people from hazardous materials.

What is a hazardous material? A hazardous material is any material that when handled may cause health problems, fire, explosion, or may react with other materials in an undesired way. Examples may include poisons, insecticides, herbicides, certain veterinary supplies, motor fuels, cleaning solvents, painting supplies, bonding agents, catalytic agents, and many more. Those given as examples here are often found on farms. Literally hundreds of hazardous materials have been identified, industry wide.

Hazardous materials safety programs should include as a minimum:

- An extensive inventory of hazardous chemicals/materials on site.
- A means of disposing of items which are unidentifiable or no longer of use.
- Making sure remaining hazardous items are labeled and a material safety data sheet (MSDS) is on file for each.
- Training all employees regarding all hazardous materials on hand as to the health hazard(s), proper handling, storage, their labels, and MSDSs.
- Maintain a notebook containing all existing MSDS sheets — to which new sheets will be added as substances requiring them are purchased.
- The MSDS notebook must be available to any employee who wants to see it.
- Make a brief written document discussing the above items, which becomes the farm's hazardous materials plan.

Farms subject to OSHA regulations are required to do these things, plus more. The above plan may seem like a tall order. It isn't. Once established, it will give the proprietor a handle on his hazardous materials, and could save workers from suffering or even death.

The first step in any hazardous materials safety plan is an inventory of what is on the site at the present time. What may be found is often surprising. Do not look only in the obvious places; look on top of cabinets, under things and behind things. Look for out-of-the-way storage areas that may be located in several buildings. Gather up and use care in handling the items. Next, attempt to identify what has been found. Look at the labels. If warnings are on the labels, the items could contain hazardous materials. What if there is no label and the contents are suspect? Get rid of it. Contact a hazardous materials disposal company. Most commercial trash haulers can provide the name of a hauler qualified to dispose of hazardous materials. Do not take it out and bury it! It could cause environmental damage, or sickness.

Set up a suitable storage area for what has been found and is going to be kept. A paint shed or similar small building might be ideal. Some materials have to be protected from freezing, so this may be a consideration. Make sure everyone connected with the operation knows where the items are stored. Next, obtain MSDS sheets on each substance you are going to keep. Chemical manufacturers usually send these sheets along with the product. Dealers should pass them on to users. The manufacturers supply these sheets which provide the chemical name plus trade names of the product, or hazardous parts of the product. The MSDS sheets also give certain personal protection information, plus special handling and Personal Protective Equipment (PPE) requirements. These sheets may be used in orienting employees with respect to the hazardous materials used on the farm. Make sure all containers have labels on them. Always store them with the labels showing, if possible. The less these containers are handled, the better.

Most users trust the information in the MSDS. OSHA law states that if a user follows the guidelines provided by the MSDS, he will not be cited if it is slightly in error. Usually, manufacturers quickly and courteously provide information over the telephone when called by a user. The phone numbers are provided on the MSDS. The manufacturer does not want to causer harm to people or animals.

It is very important that unknown or unused chemicals be disposed of properly. Some of the older solvents were found to be carcinogenic, such as the early boiler cleaning solvents. Rather than getting involved in all the permits required for disposal, call a disposal company and let them handle it.

A WORD ABOUT MATERIAL SAFETY DATA SHEETS

MSDS sheets may differ greatly in appearance. Some may be several pages long. Some have colors and/or logos and are quite fancy. Others are plain. It is important to remember that all have the same section or paragraph structure. This makes it easier to find information that may be needed.

Section I is the identification section. It provides the chemical name and often gives trade names. It provides the maker's name and address, as well as emergency telephone number(s).

Section II lists the hazardous ingredients and gives the various chemicals contained in the formula. It also provides information on the exposure limits — the amount of time that a person may be safely exposed to the product.

Section III provides certain physical data, such as vapor density, i.e., evaporation rates and whether it is heavier than ambient air. Vapors heavier than air may collect in low-lying areas. This section also discusses the chemical's ability to mix with water. Odor and appearance are also discussed.

Section IV provides fire and explosion data, if applicable. It provides information on extinguishing fires fueled by the compound. It gives the flash point, which is the lowest temperature at which a liquid can evaporate and ignite if exposed to heat. It will also disclose whether the chemical produces a hazardous atmosphere when ignited.

Section V provides reactivity data — whether the chemical may react with other materials and cause fire or explosion or release harmful gases.

Section VI provides health hazard data. It discusses routes of entry into the body — inhalation, absorption, or ingestion. This section also identifies the possible long- or short-term effects the chemical may have upon people and it describes symptoms of exposure over time, such as dizziness, nausea, or unconsciousness.

Section VII provides information on preventive measures, such as safe handling information, recommended PPE, and handling and storage methods. It also provides disposal information.

Section VIII provides first aid tips, and what to do if accidentally exposed to the substance.

Section IX sets forth other information that the manufacturer considers pertinent.

It can be seen that the MSDS provides a lot of information about a given hazardous material. These sheets are very helpful in the training of people who handle chemicals.

Hazardous material storage areas should be locked. Persons who do not need to have access to them should not be allowed to handle them. News reports have included stories of children getting into hazardous materials and suffering severe illness and even death. Neighbor children and visitors must be protected.

BUILDING REPAIR

Most proprietors engage in a certain amount of repair work that falls into the carpentry category. In colder climates, building repair is often viewed as an urgent concern. This type of repair work, like many farm-related tasks, is something that is necessary but not often required. This is the case at most agricultural building sites. Workers have been heard to say, "We don't do enough of it to get good at it." This is probably true in most cases. Agricultural workers do many jobs, each of which has different hazards.

LADDERS AND SCAFFOLDING

Scaffolding is required on many repair jobs; ladders are required on most of them. Faults with ladders, like some other occasionally used tools, are easy to overlook until they are needed. Bad, broken, or poorly constructed homemade ladders still exist in some workplaces. Using a broken or a poorly maintained ladder is like using a faulty parachute. Hardly anyone would do the latter, but many do the former.

Ladders

Stepladders are available in three classes. The least sturdy and lowest priced ladder is sold for use in the apartment or home. It is called a Type III ladder, with a height limit of 6 feet. The middle grade ladder, Type II, sometimes called a commercial ladder, is a little stronger and costs a little more. Its length may be up to 12 feet. The industrial grade ladder rounds out the field. This ladder is a Type I and may be from 3 to 20 feet in length. Which ladder is best? The answer depends upon its intended use. Your supplier can discuss the specifics of all three classes.

Ladders are usually constructed of metal or hardwood. Metal ladders are not recommended for any type of electrical work, since they are good conductors of electricity. Metal ladders require less maintenance than wooden ones and are often lighter in weight. Always inspect ladders before use. Always store ladders where they are protected from the elements and from damage. Never use a ladder with missing parts or broken rungs. Use ladders with the proper feet for the surface upon which they are resting.

Someone should steady a ladder while someone else is climbing. Ladders need to be tied off at the top to prevent their falling. Make sure extension ladders have proper locks to keep them from telescoping when in use. Tie off the extending rope as a backup. If a worker is going to move from the ladder to the top of a structure, the ladder should extend at least 3 feet above the eave line. Steps and rungs should be free of grease to provide a safe footing. A nonslip material should be applied to the bottom few steps.

Ladders placed at too steep an angle are dangerous to climb. A good rule to follow is to place a ladder one-fourth its length from the vertical face. For reference purposes, OSHA regulations that deal with portable wood and metal ladders are 29 CFR 1910.25 and 29 CFR 1910.26, respectively. Proprietors may wish to obtain a copy of the 29 CFR 1910 series of regulations. These may be purchased at government bookstores or from the Government Printing Office, Washington, D.C. Sometimes they are available free through local OSHA offices.

Scaffolding

Since farm people do not usually do an extensive amount of building or remodeling, scaffolding may not be available for immediate use. Many accidents have arisen from the use of improper scaffolding. Many unsafe substitutes for scaffolding are still seen today. People have been known to set a stack of 55-gallon drums along a wall, one on top of another, put a couple of planks across the top, and use this for a scaffold. Others have foolishly used a pair of stepladders with a plank laid from the step of one to the step of the other. Sometimes, bracket-type scaffold braces are nailed to wall uprights and planks placed on them to be used for high work.

Some of these temporary measures are dangerous and should not be used. It would be far safer to rent quality scaffolding from a firm in that business. Some cost is involved, but the firm will normally provide advice on how much you need and how to safely erect the scaffolding. Most of them have a delivery service and often will offer to help with the setup. These firms are knowledgeable when it comes to safe scaffolding. A bonus benefit is that a proper scaffold usually is faster to erect and much safer. Scaffolds should have guard rails and toe boards to help prevent falls and to keep tools and equipment from falling. Safety harnesses and a means for tying off should be provided for work on elevations such as roofs.

HAND TOOL SAFETY

Severe injuries have resulted from the improper use of hand tools, and from using the wrong tool for a job. In a farming operation, hand tools are used a great deal, and most workers are familiar with their uses. It is, however, necessary to review hand tool safety with workers from time to time. To help do the job correctly and safely, hand tools have to be utilized properly. Workers should be cautioned to use the same type and size of tool normally used for a given task. Open end and socket wrenches are better suited to removing and replacing nuts than are pliers and adjustable wrenches. Cheater bars and pipe sleeves should not be used to gain more leverage on a wrench. This practice may break the wrench and cause a back injury for the worker. Use an impact wrench on stubborn nuts, or try a loosening fluid or heat if safe to do so. Make sure that chisels are sharp and that the heads are in good condition. Chisels and drifts with mushroomed heads should be discarded. Pieces of metal may fly off them and injure people. Always use a metalworking hammer for any metal work. Carpenters' hammers are not designed for such work.

All tools with cutting edges should be kept sharp. They save time, work better and are usually safer to use. Sharp tools make hard work easier. Perhaps the old saying, "A dull man works with dull tools" contains a useful lesson.

Many tools are lost in the field. When working on a piece of equipment in cropland, tools placed on the ground are often overlooked and lost when the job is done. A canvas tool bag with a loop handle is a good investment. The worker can carry this about and when finished with a tool it can be placed back in the bag. Army aviation mechanics are often credited with the canvas bag idea. It was referred to "as my tool purse". Good hand tools are expensive. It doesn't take long to lose a lot of dollars in lost tools. Both for the sake of efficiency as well as safety, tools should be returned to their proper place in the shop when no longer needed.

Axes and mauls should be inspected carefully before use. Make sure the handles are secure. When cutting limbs off fallen trees, always limb from the opposite side of the trunk. When chopping down a tree with an axe, always inspect the upper portion of the tree for dead limbs that could fall on the person doing the cutting. Woodsmen call these limbs widow makers. Workers should be instructed to stay out of the line of fall when the tree is ready to topple.

Pry bars and crow bars are useful tools. Never use wrenches or hammer handles as pry bars. Do not use screwdrivers as chisels. The safe worker has learned to use tools for the purpose for which they were intended. Gloves provide some protection for the worker's hands.

Eye protection is a good idea when doing many hand tool jobs. In any job where a danger of flying chips or metal particles is possible, eye protection should be worn. Remember, a worker cannot wear eye protection, or any personal protective device, that is unavailable or that he/she can't find. Some workers carry their own eye protection with them at all times, but most do not. Goggles that fit over eyeglasses are often used. These may be placed in the shop and other areas where eye protection is required. The problem with community goggles is that they rarely are cleaned. Often when they are very dirty, a worker may grab a shop rag and try to clean them. This practice usually makes the problem worse. All community goggles and glasses should be cleaned periodically with warm water and a soft cloth. When eye protection goggles get very dirty, workers are reluctant to wear them. If a lot of eye protection is used, an eyewear cleaning station would perhaps be a good investment. These are available in first aid supply stores.

Some feel that regular eyeglasses provide adequate protection from flying objects. This is probably untrue. Sometimes optometrists will install hardened lenses in eyeglasses, but these are not classified as safety glasses. Prescription safety glasses are available for those who do not wish to use goggles or face shields. Nonprescription safety glasses are available as well.

PLUMBING AND ELECTRICAL MAINTENANCE

Although many proprietors and workers are able to do a given amount of plumbing and electrical maintenance, this work is perhaps best left to professionals. Electrical repairs should not be attempted by anyone who is not qualified to do electrical work. Improper electrical work may lead to electrocution, fire, and the destruction of electrical equipment. Farm neighborhoods may have a qualified electrician in the mutual aid group that can be called upon for emergencies. If electrical work is attempted, always de-energize the line on which the work is to be done. Then test for current using a test instrument. Some machines or processes may be fed from two or more sources. Make sure the item to be worked upon is completely de-energized. OSHA requires the locking-out of energy sources so that it is not possible for someone to energize equipment while others are working on it. The OSHA reference for plumbing and electrical maintenance is 29 CFR 1910.147.

WELDING, CUTTING, AND BRAZING SAFETY

Several hazards are common to welding, cutting, and brazing work. Among these are fire, eye injury, electrocution, and other forms of damage. Farm workers occasionally engage in these operations, but like many jobs on the farm, workers do not do them every day and may therefore forget the precautions that should be taken. Oxygen-acetylene welding jobs may be performed. This job utilizes tanks for the gases and an open flame. Oxygen and acetylene tanks should not be stored together for fire prevention reasons. When being transported or stored, these tanks should have the safety caps in place, be stored in an upright position, and be secured. Often, these tanks are mounted side by side on a welding cart that allows them to be moved to the welding site. These tanks should be firmly secured to the cart. Valves should be turned off when not in use.

Eye protection should be worn when welding. Darkened safety-type welding lenses are used to protect the eyes of the welder from flying slag or sparks. Welders also should wear flame resistant clothing that covers exposed skin and flame-resistant gloves to protect the hands.

Special care should be taken when welding near flammable materials. It is a good idea to have fire protection available, such as a water source and fire-extinguishing equipment. A fire watch may be necessary when welding in barns or other buildings with flammable materials stored inside. After the job is complete, the fire watch should continue for at least 30 minutes to spot and extinguish late-starting fires. Fire prevention is the key. Most farms do not have equipment necessary to fight structural fires and fire departments may be miles away.

Welding generates a lot of heat. Metal usually conducts heat well. Many workers have been burned while picking up a piece of metal too soon after it was welded. Cutting torches produce more heat than welding torches. Fire is a greater danger when cutting. Slag that falls from a cut may burn through a wooden floor. If possible, utilize a cutting mat made from flame-resistant material under the object being cut.

Electrical welding has hazards similar to the oxygen-acetylene method, plus one more. It is possible for a welder to be electrocuted if the welding current passes through his body on the way to ground. Ground cables/clamps should be firmly attached to the piece being welded before striking the arc. Ground connections on welding tables must be checked to make sure they have not been severed or loosened. Arc welding produces extremely bright light. Welders must wear a welding hood with a suitable lens that will protect their eyes from the intense light. Welders should instruct others in the area to not look at the arc. Portable welding screens should be set up to protect people working nearby.

Brazing has hazards similar to welding. Another common welding, cutting, and brazing hazard is breathing air that may contain harmful fumes. Fumes are metal particles that are suspended in the air. Some heavy metals when being welded produce fumes which can cause lung and other physical problems. It is impossible to weld, cut, or braze without being exposed to a certain amount of fumes. Some things can be done, however, to minimize the exposure of the welder and others. If welding outside, the welder should weld downwind from his body. Fans may be used outdoors as well as indoors to dissipate the fumes. The fire watch must be alert for sparks being blown by the fan into flammable materials. Those near the welder should keep well back.

Obviously, fume exposure to a farmworker who welds occasionally is far less than that to a welder who performs this work 8 hours every day. Some people, however, have far less physical tolerance than do others. For this reason it is always desirable to do it the safe way. The OSHA reference for welding, cutting, and brazing is 29 CFR 1910.252.

CONFINED SPACE SAFETY

Confined space — *Any space not designed for human occupancy, has limited means of egress, and may contain oxygen deficient or other harmful atmospheres.*

A farmstead may contain many confined spaces. Some which quickly come to mind are grain bins, silos, crawl spaces, attics, water supply tanks, manure pits, cisterns, and excavations. The OSHA regulation governing those subject to the law is 29 CFR 1910.146. Each year many people are killed or injured as a result of working in a confined space.

Sometimes it is difficult to tell what space is a confined space. The definition just described should help resolve this question. For example, consider an attic. One ventilated attic has a pull-down stairway, flooring to walk on, and electric lights. According to the definition this attic is probably not a confined space, even though humans may not occupy it very often. Another attic has no stairway. The only way to get into it is by climbing a ladder and entering it through a small opening. No visible means of ventilation or lights or windows, and very little flooring exists. This attic is probably a confined space. It is important to be able to recognize a confined space because workers and others should be warned of the dangers of entering it.

Proprietors should make a list of confined spaces on the premises. These spaces should be marked with signs to inform workers of their presence and to let them know that certain procedures are necessary prior to entering. Farms that are subject to OSHA regulations are required to formalize their confined space entry plan and utilize a written permit system for some confined spaces. Those not subject to OSHA would do well to at least comply with the spirit of the law for safety reasons. The proprietor should train his/her workers in confined space safety.

As a minimum, some safety procedures should be followed, such as:

- Signs should designate all confined spaces.
- No person should enter a confined space without informing others.
- If work is to be performed in a confined space, it should be ventilated if there is any doubt about the atmosphere.
- With a worker inside, another worker should stand watch outside and be in communication with the inside worker.
- If access is difficult, the worker inside should wear a safety harness attached to a rescue rope that runs from the site to the outside.
- At the first sign of physical distress, the inside worker should notify the watch person and exit the space.
- The atmosphere in suspect spaces should be checked for sufficient oxygen, such as in oxygen-limiting silos.
- Phone numbers of rescue services should be kept handy.

It is desirable to keep a checklist for safe confined-space entry. Measurement of the oxygen content in a confined space requires instrumentation. This instrumentation is expensive and requires calibration prior to use. Perhaps the mutual aid group has an instrument available. Another means of providing oxygen for a confined space involves blowing air into it. A blower fitted with flexible air ducting usually accomplishes this. Shove the air duct well inside the space and run the blower for an extended period of time. This procedure increases the likelihood of having sufficient oxygen in the confined-space atmosphere, but the only way to be sure is to measure it. Ventilation is a key factor in replacing bad air. All practical means should be used to achieve air exchange. No person should enter a confined space if insufficient oxygen exists. Sometimes, safe entry can only be achieved through the use of self-contained breathing apparatus (SCBA) or airline respirators.

The watch person is very important for confined space safety. He/she should be in constant communication with the worker(s) inside. In case of trouble, the watch person should call for help and may try to assist the worker(s) by use of the rescue rope. The watch person should not enter the confined space until other help arrives, since he/she also might be overcome. Persons rescued from a confined space should quickly and safely be moved to fresh air. If a rescue service is on the scene, follow their directions. Confined space safety is an important consideration for proprietors. Most confined spaces may be worked in safely if proper precautions and procedures are followed.

PEOPLE TRAPS

People may be trapped in confined spaces as well as in other areas such as excavation cave-ins and grain bins. Some soils are less stable than others. For safety purposes, all trenches should be sloped or shored if people are going to work in them. Rain can cause some soils to slip and/or cave in. People should only be in trenches when doing necessary work. Children should be kept away from excavations and should not play in grain bins.

Individuals have been known to suffocate in grain bins. During harvest time, when switching grain elevator spouts it is sometimes necessary for a worker to enter a grain bin. First of all, it is hard to see in a bin while it is being filled. Normally there is a lot of grain dust in the air. Some elevators move huge quantities of grain in a relatively short time. If a worker inside a bin were to bump his/her head and fall down, he/she could be covered with grain and suffocate within a few minutes. Elevators should be stopped when a person is in the bin.

Some grains are a bit like quicksands. Flax is a grain that has been known to envelop people. Grain that is very dry is more dangerous in this regard. A watch person should be available whenever a worker has to spend time in a grain bin.

Silo cave-ins also may trap and suffocate a person. Some silos without a working unloader may have silage stuck to the walls that may extend up for several feet. This is particularly true in colder climates where silage may freeze near the walls. This buildup may suddenly loosen and fall. If a worker is in the silo at the time, there could be a fatality or serious injury. Workers should assess the situation carefully before entering a silo alone. If any doubt exists about hazards, a watch person should accompany the worker.

Proprietors should use warning signs near confined spaces to let workers know there may be danger to them. With a relatively small work group, posting signs may seem trite, but they can be lifesavers.

FUEL STORAGE AND HANDLING

Most farmsteads maintain a sizable supply of petroleum products that are used for vehicles and heating fuel. In addition, lubricating oils and greases are kept on hand. Petroleum products always pose a fire hazard and require care in handling. Gasoline, when vaporized in proper concentration can explode, causing damage and fire. A pound of gasoline has the explosive power of over $2\frac{1}{2}$ sticks of dynamite. The gasoline by itself will not explode, but mixed with air it can be quite deadly.

Like the proverbial snake, the two-step adder, gasoline is relatively harmless if it is kept securely contained so that leaks are controlled and the area around storage tanks is kept free of flammable materials.

Certain terms pertain to flammable and combustible liquids with which farm people should be familiar. Some of these are listed and defined below.

- *Auto-ignition temperature* is the temperature at which a flammable vapor-air mixture will ignite (given defined conditions) by itself.
- *Flash point* pertains to the lowest ambient temperature at which a liquid will give off sufficient vapor to form a flammable mixture.
- *Flammable limits* are the concentrations of vapors above and below which flame will not occur.
- *Volatility* is the ability of a liquid to form vapor.
- *Vapor pressure* is the pressure exerted by a liquid in pounds per square inch. (Liquid will vaporize unless it is stored under a pressure greater than its vapor pressure.)

What is a flammable liquid? A flammable liquid is one that has a flash point lower than 140°F and a vapor pressure no greater than 40 psia (pounds per square inch absolute); absolute pressure takes atmospheric pressure into consideration. Combustible liquids have a flash point between 140 and 200°F. Thus flammable liquids ignite more quickly than combustible liquids and should be handled accordingly. These definitions are provided to aid in the interpretation of data contained on the MSDS for various petroleum products.

Flame sources should be kept away from fuel storage and distribution areas. "No Smoking" signs should be placed near the fuel pumps.

Normally, fuels on farms are stored in elevated tanks with either a gravity or force feed to the vehicles (Figure 3.1). These facilities should be located away from buildings, where they can get plenty of ventilation. Placing storage tanks inside buildings is not recommended. Vegetation should be controlled under and around tanks. Dry vegetation increases fire hazards.

Petroleum products tend to defat the skin when they get on the body. Workers should wash gasoline, oils, and greases off the skin quickly and thoroughly. Handled properly, petroleum products present no severe hazards. Mishandling, however, can be catastrophic.

HAUL ROAD AND DRAINAGE SAFETY

Timely maintenance of on-site haul roads and drainage systems can result in greater overall safety and will save time and wear and tear on equipment. With today's heavier and faster-moving equipment, on-site roadway construction and maintenance is a must. Few farms have the equipment necessary to construct haul roads from scratch. Most hire contractors to do the initial grading and rough finishing. Establishing grades and placement of culverts or hardened washes usually requires some light surveying. In most cases, proprietors are familiar with drainage patterns, if not rather specific natural flow ways. For this reason, a simple land survey will suffice. In some areas, road surfacing may be done with materials found on the

Figure 3.1 Typical fuel storage facility.

owned or leased land. If this is not the case, it may be necessary to purchase gravel or other surfacing material. Some proprietors rent the heavy equipment necessary and build roads with little outside help.

Safety considerations include worker orientation prior to beginning work to cover such topics as haul load stability, load considerations regarding weights, and the capacity of equipment including trucks needed for hauling. Towing may be necessary to get loaded trucks to the construction site. Tow chains/cables should be in good condition and towing attachments placed from drawbars to frame hooks on trucks. Workers should stay clear of chains and cables during rescue tows. All workers need to be reminded to stay clear of backing vehicles. Many of the newer farm trucks have backup alarms on them that serve to warn workers that a vehicle is backing up. If haul trucks require chains to get traction, they should be installed before entering the haul road construction site. Trying to put chains on a stuck vehicle is always a dangerous undertaking.

The final grading (scrape) prior to surfacing the roadway is important. The roadway should be packed prior to the final grade. If this is done properly, the road will last for a long period of time. If done poorly, the roadway will need continuous maintenance.

Culverts and washes should be of sufficient size to handle drainage. This, too, will give greater life to the haul road. Haul road construction and maintenance involve a cash outlay but this may be totally offset by the extended life of machines and equipment that have to move in and out of the area. Serviceable haul roads save considerable time.

HOUSEKEEPING (ORDER) ON THE FARM SITE

Good order is perhaps the first step in managing a workplace. Order in the workplace is a newer concept of what has traditionally been called housekeeping. Housekeeping's definition is "A place for everything and everything in its place." The definition of order is a bit broader: "A workplace is in order when there are no unnecessary things about, and when all necessary things are in their proper places." This definition of order is attributed to Frank E. Bird, Jr., an internationally known safety expert. Order is a first step in having a safe workplace. According to the above definitions, it would be possible to have good housekeeping, but not order.

Equipment, when not in use, should be in the equipment yard/area. Equipment should not be left in the field. Located away from the yard, it is awkward to inspect, adjust, and maintain. Machinery should not be left where cattle can mill around it. Animals often get caught in machinery and could be badly injured. Gates have to be opened and closed when retrieving a piece of equipment from an area containing animals. It is much more economical in terms of time and effort to have machinery stored in order in the yard, or in machine sheds. It is much more secure there as well.

A lot of time is saved if everything is in its proper place. Work moves more smoothly and frustrations, usually present when tools and equipment can't be found, are avoided. Disorder breaks routines, causes confusion, and does not serve the safety effort.

Control of vegetation around buildings reduces the hazard of fire. It also discourages wildlife from nesting in the barnyard or dooryard.

Finally, order makes equipment and facilities easier to maintain and repair.

PEST CONTROL

Farmsteads attract all manner of insects, rodents, and other types of wildlife. Insect control starts with good housekeeping. Good drainage, which eliminates much standing water, destroys breeding habitat for many insects. Sprays and dusts applied in accordance with the manufacturer's specifications will also reduce insect infestation. Sprays and rub ropes containing insecticides that are not harmful to farm animals may be used to protect them from flies and other pests. It is important that such elements be used and applied according to the instructions from the maker. It is also very important that employees know the properties of such insecticides for their own personal protection. Although workers may be subject to rather limited exposure, their safety must always be considered.

Respirators, skin creams, and other protective measures may be indicated for those engaged in insect control procedures. Workers' clothing should keep sprays and dusts from getting on the skin. Clothing should be changed and the soiled clothing cleaned or washed after insecticide applications are completed. Workers should take care to spray downwind, if possible, while applying the chemicals.

Many types of insecticides are on the market. Using the proper one(s) will do a better job and will probably be cheaper in the long run than engaging in a trial and error method of treatment. County agricultural agents and suppliers may be able to recommend the best products for the pests involved.

Rodents may be substantially controlled by farmyard cats and dogs. Rats that burrow into grain bins and silage piles may require other treatments. Like insecticides, many rat poisons are on the market. A chief safety concern when using poisons is applying the proper agent in such a way as to protect domestic animals and people. Here again, the county agricultural agent is usually a good source of information regarding rodent poisons and other related topics.

In many parts of the country rabbits were major consumers of ear corn when stored in cribs. Squirrels also ate a lot of corn if they could get to it. Today less ear corn is stored on the farm site, so loss from squirrels and rabbits may not be a problem. Most proprietors do not begrudge losing a little grain to wild animals unless the loss becomes sizable. Today's corn harvesting methods usually leave sufficient corn in the fields to satisfy some of the small animal pests. Do small wild animals carry diseases that may be harmful to humans and farm animals? Check with the veterinarian for an answer, because this varies from state to state.

SUMMARY

Hazards associated with farm buildings and grounds vary in complexity and in potentials for losses. Good housekeeping is a starting point for both building and grounds safety. When the growing season is complete and the feed livestock has been sold, many proprietors take special pains to store equipment and tools in an orderly manner. The question may be asked, "Why wait?" Work proceeds more smoothly and timely when equipment and tools are stored properly during the growing and feeding season. By maintaining order, the proprietor demonstrates to all that good housekeeping is important. Studies have shown that workers perform better and safer when the work site is maintained in an orderly condition. And it saves time!

Safety concerns around buildings and grounds require constant attention. Unsafe habits are easiest to change as soon as they are exhibited. An old analogy states that bad habits start out like cobwebs, but if allowed to continue they turn into cables. It is relatively easy to knock down cobwebs, but cables are a problem!

As is true in other occupations, the proprietor should set the safety example. Workers learn from example, just as they do from formal instruction. If safety equipment is required on a job, the proprietor should wear it. It is socially difficult to correct a worker for a safety violation if the proprietor is guilty of the same behavior.

Building and grounds safety may be enhanced by application of the management principle of communication. Paraphrased, this principle says, "If you have a problem or concern, tell your people about it, and ask for their help." Will they help? The answer is most often yes. On the other hand, if the proprietor does not tell people about the problem he/she has been wrestling with, they cannot help. Good communicators are usually good leaders. Leadership is as necessary in safety as it is in production.

FURTHER READING

Bever, D. L., *Safety: A Personal Focus*, 4th ed., Mosby-Year Book, St. Louis, MO, 1996.
Bird, F. E., Jr. and Germain, G. L., *Practical Loss Control Leadership*, 2nd rev. ed., International Loss Control Institute, Loganville, GA, 1992.
BOCA, *National Building Code*, 12th ed., Building Officials & Code Administrators, Country Club Hills, IL, 1993.

CHAPTER 4

Fire Safety

There are two types of fires — friendly and unfriendly.

— H.E. O'Shell, P.E.

If civilization were to lose the ability to make fire, it would gradually return to the Stone Age. Fire is often depicted as an evil thing, but it is necessary to life as we know it. It is only when fire is uncontrolled that it becomes a destroyer. Fire is a safety concern because if it is allowed to start and go its own way, the means of production and people can be destroyed. Proprietors and workers may feel that their role in fire prevention is minimal. Many feel that the main responsibility for fire prevention rests with builders and installers. This is a mistaken idea. Unfriendly fires are classified as accidental losses and are therefore part of the responsibility that everyone has for accident prevention. Fires on farms can be especially serious because professional fire fighting help may be up to an hour away. A bad fire can put a farm out of business! Though farms may have certain fire-extinguishing equipment, they are usually not equipped to fight structural fires. Builders and installers provide a first step in fire prevention by constructing facilities which are in accordance with fire codes, and making sure that electrical and heating equipment is properly wired and installed.

Even though the design of the workplace meets fire prevention standards, proprietors and workers have a responsibility to engage in activities that minimize fire risk. Like much in safety awareness, the proprietor is the key to fire prevention on the farm.

FIRE PREVENTION

The primary goal of a fire loss control program is prevention. Work to make sure that fires do not get started. Many tend to feel that knowing where the fire-extinguishing equipment is and how to use it is their first duty. This is desirable, but not the first duty. Their first duty is to do everything possible to keep fires from starting. All workers must actively support fire prevention efforts. Without everyone's help, it is nearly impossible to prevent fires. Of course, some fires may be started by natural causes, such as lightning, but most fires are caused by what people do or fail to do.

ELEMENTS NECESSARY TO START A FIRE

Three things are necessary for a fire to start:

- Fuel
- Oxygen
- Heat

These three elements make up what is commonly referred to as the fire triangle. Since plenty of fuel and air are available in most situations, it is clearly heat that starts fires. Air is made up of several elements, but enough oxygen exists in ambient air to start a fire. Heat, on the other hand, is not as available as air or fuel. Fire prevention begins with the control of heat sources around flammable materials.

Once started, fires are sustained by another element, these are the free radicals. Professionals studying the chemistry of fire found some years ago that fuel molecules combine in a given way with oxygen in what is called a branched chain reaction. During this process intermediate products are formed. These are the free radicals. These radicals, mainly unstable hydrogen groups, tend to regulate flame speed and spread. Once a fire has started the fire triangle becomes a tetrahedron. Why the technical explanation? It is perhaps unnecessary to know, but it does come into play in understanding that certain types of dry chemicals that are used to fight fires tend to interrupt the aforementioned chain reaction.

CLASSIFICATIONS OF FIRE

Workers should know the classifications of fire. These will be presented here. Additional information on these classifications as well as the nature of fire may be found by referring to publications produced by the National Fire Protective Association (NFPA). Many libraries have these available.

Fires are classified as follows:

- **Class A Fires** — These are fires that occur in ordinary combustible materials such as wood, trash, paper, and rags.
- **Class B Fires** — These fires feed on air and vapor mixtures coming from substances such as gasoline, oil, grease, paint, and other flammable solvents.
- **Class C Fires** — These are electrical fires, which usually start due to overheating of electrical equipment. *Note:* even though the initial fire may be burning electrical equipment, the fire may spread to other materials.
- **Class D Fires** — This fire involves combustible metals such as sodium and magnesium.

FIRE EXTINGUISHERS

The four classes of fires may require particular types of extinguishing agents. It is very important that the proper kind of extinguishing material be used, since the wrong kind may cause other problems and sometimes make a fire worse. Proprietors should ensure that the proper type(s) of extinguishers are available at likely fire sites. All workers should be instructed on the use of the various types. Extinguishers in untrained hands are often worthless.

Water is perhaps the best extinguisher for Class A fires. A water hose attached to a hydrant is an excellent choice, if one is in the vicinity of the fire. Running the hose to the site where hot work is being done is a good protective measure. This would enable a worker to put out a fire immediately after it started. There are several types of portable water extinguishers. All can be effective if the fire is detected early and has not spread. In colder climates, water extinguishers may freeze, so it may be advisable to fill them with a nonflammable antifreeze liquid.

Types of portable water extinguishers for use on Class A fires are

- Pressurized types — These are small tanks with a short hose, nozzle, and valve. They are partially filled with water and are air pressurized to propel the water when needed by squeezing a handle or opening a valve. This extinguisher, like most portable water types, has a range of up to 40 feet.
- Cartridge types — These are similar to the pressurized type, except that they contain a cartridge which, when activated, pressurizes the tank.
- Pump type — This extinguisher is a container of water with a pump which has to be worked in order for water to flow.
- Soda-acid type — This is not in use as much now as other types. It has a container of acid and soda inside. When the tank is turned upside down, the soda and acid mix and generate pressure, which propels the water. If old fire extinguishers are around, they are perhaps the soda-acid type.

The four portable water-filled extinguishers above are all about the same size and weight. They contain about 2.5 gallons of water and have a range of up to 40 feet. Each type requires checking and maintenance. Follow the instructions on the extinguisher both for use and for maintenance.

An extinguisher that may be used on both Class A and B fires is the foam type. Turning it upside down and pointing the nozzle at the base of the fire operates it. This extinguisher, too, has a range of about 40 feet and needs to be recharged annually.

A portable carbon dioxide extinguisher may be used on Class B and C fires. It will also control small surface fires of the class A variety, but is usually not effective on Class A blazes. To operate, pull the pin and squeeze the lever. Effective range is limited — normally not over 7 feet.

Dry chemical extinguishers may be used on Class B and C fires. They are tanks charged with dry chemical under pressurized air. To operate, pull the pin and squeeze the lever. Some dry chemical extinguishers contain an internal cartridge. Rupturing the cartridge causes the extinguisher to pressurize.

Special extinguishing agents are needed to stop Class D fires. The extinguisher often resembles a metal can full of a sort of powder. A scoop is provided in the can to place the extinguishing material on the fire. Class D fires are rare unless people are working with a combustible metal.

Some extinguishers are designated ABC, indicating that they may be used on those three classes of fires. These are particularly useful on machines used in the field, such as combines. Normally employing dry powder as an extinguishing agent, these extinguishers are probably better at fighting B and C fires rather than A fires. If the A class fire is localized and not too large, they may be very useful.

Workers need to be trained in fire extinguisher use. Give them some practice. Using extinguishers that need recharging, start a fire in a safe place and let employees extinguish it; the extinguisher is going to be recharged anyway. Sometimes mutual aid groups stage a brief fire-fighting exercise wherein most workers get a chance to use a portable fire extinguisher. During a fire is a poor time to find out if workers know how to use extinguishing equipment! Local fire departments are usually happy to orient employees. They are the experts. Let them help.

FIRE INSPECTIONS

Inspections are a key loss prevention activity in fire safety as they are in general safety. Ordinarily, fire inspections are included as a part of a more general safety-oriented inspection. This makes the best use of time and effort. Once an inspector, whether he is the proprietor or a worker, has trained himself to recognize fire hazards, the inspection may contain fire prevention topics along with any others. The key to any safety inspection is knowing what to look for and what to look at. Following is a list of items that may be observed on a fire prevention inspection, or as a part of another inspection:

- Brooms, mops, and other equipment stacked in front of circuit breakers
- Fire extinguishers located in blocked or not readily accessible areas
- Paints, solvents, and other flammable liquids in open or unmarked containers
- Doorways partially blocked by trash or equipment
- "No Smoking" signs missing
- Ground prongs broken off electrical tool or cord plugs
- Trash or debris piled in corners and under stairways
- Combustible materials stored near welding, grinding, or cutting areas
- Greasy or oily rags in flammable containers
- Evidence of smoking in nonsmoking areas
- Partially discharged fire extinguishers
- Dry weeds and combustible trash around refueling stations
- Frayed or damaged power cords
- Smells that suggest something is hot or burning
- Combustibles around portable heaters when in use
- Crop debris on machines that may come into contact with sources of heat

These are some of the things to look at and conditions to look for as a part of a fire prevention inspection. There may be others since all facilities are different and equipment may vary from one farm site to another. Be particularly watchful for repeat things. It was substandard last week, and it is the same way now. Repeat items should be brought to the attention of the proprietor. The inspector, when finding something amiss, should correct the condition then and there, if practical. No need to write it down or classify it if it can be fixed when found. If the hazard is complex and requires more help or needs permission to fix it, notify the proprietor.

FIRE IN THE FIELD

Field fires can cause injuries to people and animals, but their primary danger is to equipment and crops. It has been estimated that insurance pays out at least $10 million a year for crop and machine losses. These are insured costs. Uninsured costs may be as much or more than that figure.

Field fires are most likely to occur when vegetation is dry. Dry vegetation plus wind can cause a fire to increase in intensity and spread very quickly. Exhaust systems on vehicles and other sources of heat, such as bearings, can provide sufficient heat to start a fire. Catalytic converters on trucks and pickups are generally closer to the vegetation than those on tractors and self-propelled machines. These converters operate at very high temperatures, which may be as much as 500°F. If these converters come into contact with dry stalks and grass, a fire may be started. Workers should be mindful of this when operating in the field. Trucks and pickups should be driven on driveways and lanes rather than across the field when servicing field machines. During dry periods workers should stay alert to conditions which may cause fire and avoid them if possible. Often, fires are out of control before they are discovered.

Operators of self-propelled harvesting machines may smell smoke before they actually see it. The source must be located quickly and the fire extinguished before it damages the equipment or spreads to the cropland. A 10 pound ABC extinguisher should be located where the operator can quickly grab it, such as in the vehicle cab. A larger extinguisher may be located so that it may be easily reached from ground level. Mounting it near the cab ladder might be a good choice of location. It is also a good precaution to carry extinguishers in truck and pickup cabs when these vehicles are being used to service field machines. Extinguishers must be checked prior to anticipated use, such as harvest season. Make sure they are charged and in a useable condition.

KEY ELEMENTS OF FIRE SAFETY

As is the case with all safety work activity areas, the key to fire safety may be summed up as:

- Recognition — be alert; note exceptional things; ask questions.
- Evaluation — how does this affect organizational safety?
- Control — do something to neutralize or eliminate the hazard.

These key words apply to most safety work. These same words are usually found posted at the front of safety classrooms. These three words are usually the ones safety professionals will answer with if asked what they do for a living. In fact, are these the things that are done to attain productivity, quality, and cost control? Yes! Fire safety, like other forms of safety, is something we do — just like producing quality products at a competitive cost.

Another lesson learned in fire safety is that one has to keep at it. After a time, particularly if those doing it are doing a good job, it tends to become sort of boring. The tendency is to slack off and lose interest. This can be a grave error. Safety work is a never-ending job. It has to be done again and again — just like mowing the hay. But like mowing, when it's done, it's done. Workers may then go on to something else. Fire safety work is like that. When the fire safety inspection is done, it's done. Of course, if a fire hazard is discovered at any time, it should be corrected or reported. Many workers take pride in performing their safety duties well, just like their other work. Good for them!

FURTHER READING

Cote, R. P. E., Ed., *Life Safety Code Handbook*, National Fire Protection Association, Quincy, MA, 1994.

Laughlin, J. W., *Private Fire Protection and Detection*, International Fire Service Training Association, Oklahoma State University, Stillwater, OK, 1979.

CHAPTER **5**

Machine Safety

Today farms are much more mechanized than they were a generation or so ago. Mechanization has brought with it many more hazards than were common earlier in the agricultural setting. Today's machines, though similar in some ways to earlier ones (Figure 5.1), are larger, more complex, and require a great deal of skill to operate them safely. Early tractors were little more than horse substitutes. Many had less power than a four-horse hitch. They were less complex than today's large powerful machines.

Most harvesting machines were tractor-drawn and received their operational power through use of a bull wheel and chain which drove the moving parts. Today many machines are self-propelled with all the power sources carried onboard. Towed machinery often has power supplied to it by a power takeoff from the towing tractor. Some towed machinery has a motor onboard and relies on the tractor only for movement. Point of operation devices, such as sickles, snapping rolls, and other gathering devices are much the same as they were earlier, except they are much larger and many are designed to run at greater speeds. Today's proprietor, who is perhaps 50 years old, has seen a rapid development in farm machinery and farming techniques during his/her lifetime.

This chapter will consider the following areas of machine safety:

• Machine guarding
• General machine safety
• Tractor safety

MACHINE GUARDING

Development of guarding on most farm equipment began in the 1940s. With some exceptions, agricultural machine guarding was slower to be developed than that for factory machines since little or no legislation was required, as is mostly the case today. Larger farms, those that employ more than 10 people, fall under the jurisdiction of OSHA and have had guarding requirements since 1970 when that Act became law.

Figure 5.1 A Farmall F-12 row-crop tractor used on a 160-acre farm in the mid-1930s. (Courtesy of Adrian E. Dalen, Colton, SD.)

GUARDING PRINCIPLES

What are guards on machines supposed to do? They keep workers and workers' body parts from being injured by various machine hazards. Guards, according to law and common sense, should be durably constructed, firmly attached, and not present a hazard in and of themselves. Considered here will be two types of guarding:

- Power transmission guarding
- Point of operation guarding

Transmission guarding is the easiest to accomplish and usually requires less adjustment and maintenance than the other. Transmission guarding, as the name implies, means to cover power takeoff shafts, gear trains, and other devices that carry power from one place to another. In many instances a gear train may be completely enclosed by a guard so that it would be impossible for a person to become injured by it. Transmission shafts, such as from the tractor's power takeoff to the equipment being powered, may be covered so that a person's clothing cannot become wrapped around it.

Guards are normally designed so that they may be rather easily removed for oiling, greasing, and other maintenance. They are also rather easy to put back on. The problem is that they are sometimes not replaced and the shaft, for example, spins in the open ready to grab a piece of clothing and literally wrap a person around it! Such accidents usually result in very severe injury or death. Proprietors should make sure that all guards are in place before a machine leaves the yard. It takes very little time to replace a guard. This is time well spent. Unguarded gear trains can

Figure 5.2   In-running nip points formed by the snapping rolls on a corn-picker head.

destroy the human hand in an instant. Since power transmission guarding is relatively easy to accomplish, running without it cannot be justified. Belts and pulleys are other examples of power transmissions. Guarding them is relatively easy — just cover them up.

Some older equipment may not have full-cover transmission guarding. Retrofitting may be costly, but severe injuries are as well. When purchasing new equipment, it is advisable to make sure that all available guards are with it. Guards available as an option, but not included, cannot protect farmworkers.

Point of operation guarding is harder to achieve. The operating point(s) do necessary work, such as cutting, gathering, and lifting. If one were to enclose a mower sickle like a power shaft, it would be silly. A sickle cannot cut when it is completely covered by a guard. Of course, guards are on sickle bars but they do not cover the complete point of operation. Snapping rolls on a corn picker head are in-running nip points (Figure 5.2) and do considerable physical damage if a person gets caught in them, but they have to operate in the open so they may pull corn stalks through and remove the ear. They will also remove a forearm if one gets caught in it.

Some guards are of a proximity-limiting type. For example, the straw spreader blades on the back of a combine may have a tubular metal guard that would prevent a person from walking into the blade arc. Proximity guards also serve as reminders something is near that should be avoided. Some proprietors paint such guards a bright yellow or international orange — both easy-to-see colors. Some guards on factory machinery are electrically interlocked. The electrical connection is severed when a guard is removed and the machine will not start until the guard is replaced. Interlocks on agricultural machinery are rare.

Table saws have guards that will ride up over the piece being cut, which makes it very hard for the sawyer to get his fingers in front of the blade. Often, table saws are found with the guards removed. This should not be the case.

Many devices which of themselves are not guards may be used with guards. For example, when using a table saw a push block keeps fingers away from the saw blade. Safety clutches, sometimes found on silo filling machines, which stop the feeder if a worker leans too far in front of the chopper and fan, are examples of safety devices. In factories, large shears and presses are sometimes equipped with light senders and receptors. When a part of a person's body breaks the light beam, the machine stops. These devices would be very hard to use outdoors around agricultural machinery. Dead man switches which shut down a piece of equipment when the operator leaves his station are another example of safety devices. They are not guards. Blocks or stops that prevent a radial arm saw from being pulled out past the edge of the table are other devices. A chain which limits the forward motion of a swing saw is yet another.

Other devices designed to protect the operator from injury on power tools include the blade enclosure on band saws. This device covers the band and can be moved down to keep the point of operation as small as possible and still allow room for cutting the desired stock. Sometimes this device is left in a near full-up position that exposes several inches of the blade. This device should be adjusted for each thickness of stock being cut. Leave enough room so that the stock may be cut without binding on the device. Leaving the device too high exposes additional blade and increases the likelihood of injury.

Bench grinders have guards or shields that cover a large part of the grinding wheel. It is important that the tool rest be properly adjusted. Tool rests should be set so that they are 1/8 in. from a properly dressed wheel. A wheel is dressed when it has a flat grinding surface. Sometimes a grinding job may leave an indentation or furrow in the face of the wheel. The wheel should be dressed so that its grinding surface is once again flat. A wheel dresser is a good tool to have in the shop. The tool rest should be adjusted to 1/8 in. after the wheel dressing is completed. Eye protection should be worn when using a bench grinder. The replacement grinding wheel should be the proper type for the grinder upon which it is to be installed. The wheel should be capable of withstanding the RPM centrifugal force exerted by the maximum idle speed of the grinder motor. The maximum allowable RPM of the wheel is normally found on the end paper that is attached to a new wheel, the maximum RPM idle speed of the grinder can usually be found on the information plate on the motor. Grinding wheel shields or housings should be kept in place. Never remove these housings or attach a buffer wheel on the spindle when a grinding wheel is in place. Children should not be permitted to use a bench grinder.

Butchering is done on many farms. Mesh guards are available for the holding hand and helps avoid knife cuts. If meat is being cut with a band saw it is better not to wear the metal mesh guard, because getting a hand caught in the band saw blade while wearing such a guard usually results in a more severe injury than without it. Some grinders and saws have a brake installed that will stop the blade very quickly when the power is turned off. These devices are worthwhile safety features. Loose-fitting gloves should be avoided when working with power saws. Many chain saws have such brake devices. Some are fitted with a tip guard that precludes a rapid upward jog if the tip comes into contact with the log being cut.

Guards and devices are designed to keep people from being injured or killed when operating equipment. These should be maintained in working condition. Part of a farm's safety inspection process should concern itself with keeping all guards and devices in place and in good repair.

MACHINE AND IMPLEMENT SAFETY

This section is concerned with farm implement safety as it applies to planting, cultivating, harvesting, and related equipment. Tractors present some special safety concerns and will be covered later. A primary concern in farm implement safety is proper maintenance. Manufacturer's recommendations regarding machine mainte-nance should be followed carefully. Doing so will greatly assist in safe and efficient operation. It also tends to give equipment a longer useful life.

Machinery should not be altered without the manufacturer's permission. To do so usually invalidates warranties and may impair safe operation. Most manufacturers are open to suggestions regarding modifications that will make their products more efficient and useful. They are also concerned that unauthorized modifications may present hazards that could damage the equipment or injure operators. Always consult the manufacturer regarding modifications.

Dealers can usually put the proprietor in touch with the proper people. If prob-lems arise during operation or maintenance of equipment it is very worthwhile to contact the dealer or manufacturer. They can provide additional information that should be helpful. It is an excellent policy to review equipment manuals prior to seasonal use. Often, safety instructions are in the front of the manual. This is also an excellent pre-job meeting topic.

MAINTENANCE AND ADJUSTMENTS IN THE FIELD

It is often necessary to repair or maintain equipment while in the field. This action may produce more hazards than if done in the shop or yard. Caution workers not to work on moving or running machinery. Never attempt to clear clogs or jams while a machine is running. Many people are injured each year while engaging in this unsafe practice. Operators often fail to realize the speed at which in-running nip points, such as snapping rolls on the corn-picking head, are running. People have been badly injured poking on a jam on the snapping rolls with a stick or cornstalk. If the machine is running, the rolls may grab the stick or stalk and pull the operator's hand into the pinch point. Tests have shown that the stalk is pulled into the rollers so fast that the operator cannot release it in time to prevent injury.

Figure 5.3   High pressure hydraulic lines and pump driven from a power takeoff.

It simply pulls his hand and arm into the rollers. Snapping rollers on picker heads pull stalks through at a rate approaching 12 feet per second! Hay baler heads that appear to move rather slowly take in materials at a speed of 4 to 5 feet per second. That is a lot of distance in a fairly short time!

Loose clothing may catch in gears or belt and pulley systems and drag the operator into the moving equipment. Grease guns may slip off the fitting into a turning pulley. It must be remembered that in the field one's footing is usually poor and it is easy to lose balance and fall into moving equipment. Always shut the machine down to grease, adjust, or repair it. Many pieces of planting and harvesting equipment have a power-operated raise and lower capability. Parts such as picker heads are usually raised by hydraulic pressure. Should such pressure be lost, for whatever reason, the raised part may descend and crush the person under it. If adjustments cannot be made while the equipment is on the ground, blocks should be used as a safety measure to keep the part from falling.

Many machines make use of hydraulic hoses and couplings under pressure (Figure 5.3). Never use hydraulic hoses and fittings as hand holds when mounting or dismounting the equipment. These could come loose and injure the worker. Make sure brakes are set and/or ground-engaging equipment is lowered so the implement will not move while being worked upon. This is particularly important on hilly terrain. Uncontrolled movement of a piece of equipment could injure people and/or damage property.

Mounting and dismounting of equipment may occur many times a day. The floor level in combine cabs is often 6 feet or more from the ground. Mud buildup on shoes and ladders may cause slips and falls. Wet and muddy conditions require additional care on the part of operators. Ladders should be kept free of dirt and mud and grab bars should always be used. Falls from agricultural equipment, such as combines, usually causes more injuries than getting caught in them. Equipment operators should always wear serviceable shoes with nonslip soles.

In-field collisions may occur when turning or bringing haul trucks or wagons alongside combines. Some operators try to unload on the run. Even under the best of conditions this is a dangerous practice. Accidental damage to operating equipment is expensive. It is far safer to stop the combine and then move the haul vehicle into position. Keep hands and feet out of hoppers during off-loading operations. Be particularly watchful for fire during off-loading. If a truck is being used and field conditions are dry, its catalytic converter could start a blaze by coming into contact with stalks, leaves, or stubble. The area under a truck should be carefully checked for traces of fire when it moves away.

Many accidents happen when vehicles are backing up. What is the best way to prevent backing up accidents? Don't back up! This answer may appear trite, but if operators do a bit of prior planning, they can eliminate almost all need to back up. A bit of planning may reduce the need to back up at storage bins or other off-loading operations. If an operator has to back up, he/she should do it right away — before other equipment or people can get behind the vehicle. Blind spots to the rear exist with most self-propelled farm equipment. If possible, have someone be the guide when backing up must occur.

The safe operation of field equipment requires skill and alertness. Equipment operators should be cautioned to stay clear of embankments or ditches. Be especially alert if operating at night. Equipment rollovers carry a severe injury hazard as well as a sizable damage potential. Difficult traction conditions can also cause problems. How does one safely deal with a combine stuck in the mud? A similar question was asked of a flight instructor pilot. The student asked the instructor how to get a particularly tricky airplane out of a graveyard spiral. The answer was, "Don't get it into one in the first place!" Perhaps that answer fits the stuck combine question. Operators should stay out of difficult traction conditions if at all possible.

Sometimes, errors in judgment can occur and a piece of equipment becomes stuck. When this happens operators should call on the help of the proprietor and other workers and get the vehicle back on solid footing with the least amount of damage. Towing equipment such as chains, ropes, and/or nylon tugs should be in good repair. Check the equipment manual for tips on where to attach towing equipment.

Do not yield to the temptation to attach logs, etc. to drive wheels in an effort to increase traction. This behavior is an accident waiting to happen. If two vehicles are used to pull the stuck vehicle, use a tandem hitch as opposed to a vehicles in line hitch. The latter will probably break the rope, chain, or cable attached to the stuck vehicle. Many operators prefer nylon tow straps to chains and cables. They are easier to handle, weigh less, and most have a breaking strain of over 12,000 pounds. Always check exact specifications before investing in towing equipment.

No riders! Allowing extra people to ride on operating equipment is flirting with accidental loss. An old safety saying, "No riders unless there is a seat for them," probably deserves repeating.

GRAIN DRYING OPERATIONS

Drying of grain is a seasonal function. Since drying begins with the harvest, these systems must be up and running at that time. Maintenance and safety checks should be made prior to the beginning of harvest in order to perform this work in a manner that is not pressed for time. After harvesting starts is not a good time to discover that the drying system needs cleaning and that proper maintenance has not been performed. Unexpected breakdowns and safety problems are costly in many ways, often including having to interrupt the harvesting operations. Some of the items that may require attention are

- Bearings, belts, and safety devices
- Removing debris such as old grain, dirt, and bird and rodent nests
- Fuel supply, gas lines, and wiring
- Calibration of the moisture detector
- Lubrication, including gearbox oil levels

Orient workers, particularly those who may have never operated a dryer. Go through the operation, step by step, including safety precautions. When the dryer is ready to run, it is a good idea to test it. Start it up and check burners and moving parts to ensure that all are working properly. This will greatly assist in precluding dryer problems as the harvest begins.

GRAIN AUGER SAFETY

Few farm implements have reduced hard and dusty labor conditions as much as has the grain auger. Those who labored many years shoveling grain from bins appreciate this particular piece of equipment. Augers have a rotating blade that moves grain with a sort of corkscrew action (Figure 5.4). Their sizes run from 4 inches to some 14 inches in diameter and from 20 to 100 feet in length. Most are powered by an onboard engine that makes them a very portable as well as useful tool. In spite of their usefulness, these machines are considered by many to be among the most dangerous on the farm. The corkscrew mechanism creates a very efficient shearing action should someone get a body part into the feeder end. Some of these auger elevators have a bird cage guard over the feeder portion but they usually have a fairly large mesh so as not to impede grain from entering. On most units it is still possible to poke a hand or finger into the shearing action even with a guard in place.

Children should never be allowed in the area where an auger is in operation, and workers should receive an orientation or training session prior to operating the machine for the first time. Never allow inexperienced operators to use an auger. Safe operation of augers is a good topic for preseason planning and get-togethers. Check the manufacturer's manual for other safety information.

Extreme care should be taken when setting up and using an auger around overhead electric lines. They can very easily become entangled with overhead wires. On roadways, augers are hard for motorists to see. If possible, they should be moved over public roads only during daylight hours and during times of light traffic.

Figure 5.4 A tyical grain auger. Note the unguarded pick-up end.

TRACTOR SAFETY

A review of many sources of information regarding injuries on farms shows tractors to be amongst the top three causes. Other machinery is sometimes shown as being involved in more injuries than tractors; sometimes it is the other way round. A few statistics show cattle among the top three agents involved in farm injuries. The geographic area of the country from which the information was gathered probably accounts for the differences. In any case, tractors rank high as agents of injury.

A major event with respect to tractors involving death and/or serious injury is tipping. Why do tractors roll over? A look at the tractor's physical characteristics will suggest answers to the question.

- They normally have a high center of gravity.
- They normally have three-point suspension.

Automobiles are more stable than tractors. Their center of gravity is rather low, and they have four-point suspension. Almost all tractors, especially those in the row-crop classification, have three-point suspension. Some may look like they have four if they have wide wheels in front, but if the mounting is checked it will usually reveal that the wheels supporting the front of the tractor are attached at a single strength point. Three-point suspension, by its very nature is less stable than four. Add a high center of gravity to three-point suspension and one has a vehicle that will roll over rather easily. The above description includes most row-crop tractors.

An OSHA standard applies to agricultural tractors. This standard requires that all tractors manufactured after October 25, 1976 be equipped with Roll Over Protection Structures (ROPS). The standard also states that a seat belt be installed on every tractor which has ROPS. Homemade ROPS are not recommended, since they may not meet strength requirements and may provide a false sense of security to operators. This standard also has a section that requires employees to receive instruction that references nine safe practices for tractor operation. If followed, it is hoped that such instruction will reduce the likelihood of tractor upsets. The nine safe operating practices included in the standard are

- **Fasten your seatbelt if the tractor has a ROPS.** Older tractors, not equipped with a ROPS, should not be equipped with a seat belt.
- **Where possible, avoid operating the tractor near ditches, embankments, and holes.** "Stay as far back from the edge as the ditch is deep" is a safety practice. If the operator happens to drive off a steep slope, he/she should throttle down and steer down the slope to reduce overturn probability. If the operator allows the tractor to wander off the road and head into a ditch, he/she should throttle back and gradually steer back onto the road.
- **Reduce speed when turning, crossing slopes, and on rough, slick, or muddy surfaces**. Tractor upsets are more likely to occur on slopes, especially if a front-end loader is used. Keep all loads low while moving. Keep brakes locked together (if so equipped) for road use. When tractors bounce, they lose stability. When turning, centrifugal force at 15 mph is nine times as high as it is at 5 mph. Slow down and make as wide a turn as possible.
- **Stay off slopes too steep for safe operation**. Go down slopes in the same gear needed to go up them. If it is necessary to go up steep slopes, back up them if possible. When working on slopes always keep the wheels set as wide as practical.
- **Watch where you are going, especially at row ends, on roads, and around trees**. When trailing implements in fields with obstructions, be mindful of where the outmost parts of the implement are at all times. Use care at field ends so as not to turn into part of the trailing hitch. It may be advisable to reduce speed on row and field ends.
- **Do not permit others to ride**. Make it a rule: No seat! No rider! Children should never be allowed to ride on a tractor.
- **Operate the tractor smoothly — no jerky turns, starts, or stops.** Engage the clutch slightly before increasing engine speed when starting on slopes. Sudden release of the clutch with full throttle could cause the front of the tractor to raise, particularly when headed up a slope. Locked wheel turns by using one brake is hazardous.
- **Hitch only to the drawbar and hitch points recommended by tractor manufacturers.** Loads attached to the rear axle or to the rear lift arms can cause backward flips. Check the operator's manual for allowable hitch points on each tractor.
- **When the tractor is stopped, set brakes securely and use park lock if available.** Turn off the tractor engine and set the brakes before dismounting. Place blocks at rear wheels if on a slope. On manual transmissions, place the gearshift in the lowest forward or reverse gear.

This section of the Federal ROPS Standard requires that employees must be given instructions when hired and at least once a year thereafter, regardless of the age of the tractor being operated. It is recommended that all tractor operators review the nine safety practices and follow them. It is further recommended that each employee sign a dated statement that he/she has received instruction on the nine safe operation practices when instruction is complete. A suggested statement for use with operators is

"I, _____, have read or had the nine safe tractor operating practices explained to me. I understand the dangers associated with operating a tractor if I do not follow these and other safe operating practices."

Signed: _____

Date: _____

As mentioned earlier, only farms with 11 or more employees are subject to OSHA enforcement proceedings. Since OSHA standards are considered minimums, it would serve proprietors well to be mindful of them as they apply to the farm workplace.

Other safe practices that may apply to given situations are

- Make sure the tractor brakes are properly adjusted.
- Check tires for damage and inflation level prior to operation.
- Do not pull wagons down steep grades if the wagon weighs more than the tractor.
- When operating with a front-end loader, keep the bucket or other load as near to the ground as practical.
- Keep a good muffler on the tractor.
- Keep the battery and starting system in good repair.
- When jump-starting, make sure the tractor is out of gear.

OPERATING AT NIGHT AND ON PUBLIC ROADS

Tractors and other self-propelled farm implements, when being operated on public roads at night, must have the same headlamps and taillamps that are required for other vehicles. Two headlights and at least one taillight are required. Vehicles sold with two taillamps must have both in working order. Taillamps may be mounted from 20 to 72 inches above ground level.

Farm wagons and other towed implements must have either two red taillamps or two red reflectors on the rear. Pre-1984 implements may have only one red lamp or two red reflectors. All equipment moved on public roads must display the Slow Moving Vehicle (SMV) emblem at all times. Animal-drawn vehicles must have one white lamp visible from the front and two red lamps at the rear, marking the width of the conveyance.

Rear-facing white working lights must never be used on public roads at night. State law normally prohibits these. The lights may blind drivers approaching from the rear and cause confusion as to what may be in the roadway.

Amber flashers used on many types of commercial vehicles are not required on farm vehicles, although they may be a good idea. They improve visibility and may inspire caution by approaching drivers. The lights provide another opportunity for others to see the farm equipment on the road. Safety would be well served by adding flashers to farm equipment.

By law, farm vehicles do not require turn signals. But like flashers, it would be a good idea to have them available. Of course, no lighting system will do any good if it is not used. Lights should be turned on at dusk. Many proprietors require the lights be on any time one of their farm vehicles is on a public road. Since farm vehicles travel slower than most other traffic, being seen by others is a safety factor.

Vehicle trains have special requirements in most states. A train is usually defined as a tractor towing at least two units. No more than two vehicles may be pulled behind a tractor without a permit, although this may vary somewhat from state to state. If questions arise about specific state requirements, check with the state motor vehicle department. Many states limit the total length of a farm vehicle train; 60 feet is commonly used. Sides of vehicles in a train should have a red light or reflector installed. During the day, 12-inch-square red flags should be displayed on the rear vehicle, one at each corner.

SMV emblem use applies whether operating during the day or night. All vehicles, including animal-drawn types that travel at speeds less than 25 mph, must have the SMV emblem in place. If the tractor's emblem is not blocked from view from the rear, the towed implement need not have one. It is advisable, however, to have such an emblem on all vehicles that travel on public roads. Emblems should be kept clean. If they are faded, they should be replaced. These emblems should be mounted with the lower edge between 2 and 6 feet from the ground and either be centered or to the left center of the equipment. SMV emblems should not be used for other purposes, such as to mark driveways or mail boxes. In some states a left rear-mounted flashing amber light may be used in lieu of the SMV, but it is a good practice to have SMV emblems installed.

SUMMARY

Machine guarding and farm vehicle safety are primary cares of proprietors. Leadership by the boss in safety matters cannot be overemphasized. When the proprietor shows a concern for safety, others tend to share that concern. Many hazards can be avoided by planning activities and maintaining equipment properly. Proprietors should share their concerns and ask workers for their help in solving safety problems. The old saying, " None of us are as smart as all of us," provides a guide for problem solving. The time to deal with loss situations is before they arise. Although little spare time is available on today's farm, the period between production seasons is the best time to anticipate trouble and to plan to either eliminate or deal with it.

FURTHER READING

Jepsen, H. and Bean, T. L., Tractor Tips, Ohio State University Factsheet AEX-993-96, Columbus, November, 1998.

Miller, L., Ed., ISU Study Shows Need for Grain Auger Safety, Iowa State University Extension, Ames, News Release 8-28-98. (*http://www.ae.iastate*.edu/safety/auger.html)

Purschwitz, M., Questions and Answers on Lighting and Marking of Farm Equipment on Public Roads, University of Wisconsin Extension, Agricultural Engineering, Madison, 1998.

Steel, S., The Plain Facts About Tractor Safety, National Education Center For Agricultural Safety, University of Iowa, Peosta, 1997.

CHAPTER 6

Environmental Safety on the Farm

Environmental safety covers a wide range of topics. Some of these have affected farmworkers in the same ways for many years, such as extremes of heat and cold, noise, animal and insect pests, poisonous plants, and others. Newer hazards resulting from the use of chemicals in agriculture are only now being appreciated for the effects they may have on humans and livestock. An old chemical industry axiom has a place in chemical safety on the farm: "All chemicals may be dangerous, but all chemicals can be handled safely." Animal ailments that may affect humans, the zoonotic diseases, will be covered in Chapter 7 dealing with animal safety.

Many farm people have been diagnosed with various types of cancer. While no single cause for this can be isolated, farmworkers come into contact with sunlight and various types of chemicals in the course of their job. William S. Velasquez, M.D., FACP, Director of the Hematology/Oncology Clinic at the University of Texas Medical Branch, Galveston, reports the risk of developing lymphoma has increased in certain populations with high exposure to herbicides and insecticides.

ULTRAVIOLET RADIATION

Farm people have a high risk of developing skin cancer. This is because they spend a lot of time outdoors and are therefore exposed to a lot of sunlight. A portion of sunlight is ultraviolet radiation and this, according to medical people, is the cause of many types of skin cancer. No firm figures are available that give the numbers of skin cancer cases one may expect in a farm population, however, it is estimated that over 500,000 people in the U.S. develop skin cancer each year. More skin cancer cases are diagnosed each year than any other type. The rapid rise in numbers for this type cancer has been attributed to a thinning of the ozone layer that surrounds the earth. In recent years this layer is reported to have been less dense than ever before. The cause of ozone depletion has been blamed on certain hydrocarbons that are and have been released into the atmosphere. Many of these offending chemicals have now been banned and replacements for them have, in many cases, been developed.

Staying out of the sun is perhaps the best way to cut the risk for developing skin cancer. This is impossible for farmworkers. In order to protect themselves, workers should wear proper clothing and, as an extra care, use sun-blocking lotions. Sunscreen products carry a number that is called a Sun Protection Factor (SPF). Lotions with a SPF of 15 are often recommended. Always apply the lotion in accordance with the manufacturer's instructions. Normally, sunscreens will protect a person for about $2\frac{1}{2}$ hours. Sunscreen should be applied about 30 minutes before going outside, unless directions on the container state differently. This allows time for the lotion to react with skin. If someone will be in the sun for long periods of time, sunscreens should be reapplied. Water-resistant products are available for swimmers. Direct sunlight is not necessary for exposure; one may be exposed to ultraviolet rays on a cloudy day.

Probably the best way for farm people to protect themselves from the sun is to wear clothing that covers the skin, plus sunscreen. Farm children need protection too. Health officials ask people to watch for signs of skin cancer. Generally, any change in the skin that cannot be attributed to something specific and lasts for more than a few weeks should be reported to a physician. Some medicines increase sensitivity to the sunlight. Often this information is noted on the label. If questions arise in this regard, check with the pharmacist.

MANURE AND SILO GASES

Life-threatening risks are associated with confined spaces. The atmosphere in a confined space is always suspect until proven reliable. Manure gases are always present in enclosed manure pits. These gases may be carbon dioxide, hydrogen sulfide, methane, or ammonia. All are hazardous and may be fatal in high concentrations. Any agitation of the stored manure releases large amounts of these gases. If someone must go into or work inside a manure pit, it should be treated like any other confined space, and a watch person should accompany the worker. A respirator might be needed. Always check with a reputable safety supplier for the proper respirator for manure pit gases.

Silo gases normally accumulate early — in the first few weeks after filling. The primary silo gas is nitrogen dioxide. This gas is a highly toxic agent that should begin to dissipate after four weeks. Remember that oxygen-limiting silos do not have enough oxygen to support life in normal cases. If someone must enter the silo within three weeks of filling, or at any time with oxygen-limiting types, the atmosphere must be measured for oxygen and nitrogen dioxide concentrations. The mutual aid group might well share the cost of this instrumentation, since most farms would be able to use the equipment.

PESTICIDES

Many instances occur wherein people may come into contact with pesticides. Among these are the handling, storage, mixing, and application of pesticides, or because of equipment malfunctions. Disposal activities present another opportunity for exposure.

Some of the pesticides developed and used in earlier times have been banned, such as chlordane, DDT, and silvex. These may still be found at some farms. They should not be used and should be properly disposed of. Never dump pesticides. Severe health and liability issues could arise. Rather, have them taken to a hazardous material site by a certified disposal company. Refuse collection people may be able to help, or they can usually recommend an organization that does this type of work.

It has been reported that workers are reluctant to take exposure to pesticides seriously. While it may be true that some people can tolerate rather high doses of some of these chemicals without developing symptoms, they can nevertheless be very harmful. It is not worth taking a chance with such products. Use the data provided in the MSDS as a guide for safe handling, application, and clean up. Being exposed for only a small portion of the working year is not an excuse for disregarding the risk. Pesticides may have a chronic as well as an immediate effect.

Any pesticide use subjects personnel and animals to some degree of hazard. Pertinent questions come to mind. How great is the need to use them? Would using less be feasible? If less could be used, or if they were not applied for a year, would the farming operation suffer unduly? If pesticides are not used there is no contamination exposure. The proprietor and his/her people can best answer these questions.

Always read the labels and check the MSDS on all pesticides purchased for use. To eliminate exposures during storage of these items, only sufficient quantities should be ordered for a given application. As a general rule, always wear respirators and protective clothing when handling pesticides. For oil-based pesticides use a filter with an R or a P designation. The R filters last for 8 hours. The P models last a little longer, but the recommendations on the accompanying instructions should govern. Filters come in 95, 99, and 100% efficiencies. The 95% models are sufficient for most applications. The 100% filters must be used whenever HEPA filters are required.

When handling pesticide concentrates and other very toxic substances a worker should wear goggles and a respirator plus liquid-proof clothing. A wide-brimmed hat made of waterproof material is also a good idea. Rubber gloves and boots with the pant leg on the outside complete the protective clothing. Clean clothing, respirators, and goggles each day. Never wash contaminated clothing with the family laundry. Dry and store protective clothing and devices away from other clothing. Pesticides should be stored in a locked area, preferably apart from other inhabited buildings. Children should be kept away from such substances.

PESTICIDE TOXICITY

When pesticides enter the body, they are in position to have some systemic effect. Pesticides may enter the body by three major means:

- Inhalation
- Ingestion
- Absorption

Since no person would knowingly ingest a pesticide and since the chance of absorbing a harmful dose through the digestive system is relatively small, inhalation is the most probable route of body entry. Because pesticides may contaminate food and drink, they should not come into contact with the chemical. Food products should not be allowed where pesticides are being handled, mixed, or applied. Since the human body is mostly water, pesticides that are water soluble may tend to be absorbed through the skin. Protective clothing while handling pesticides guards against absorption. Breaks or cuts in the skin may be another route of entry. A suitable respirator will prevent inhalation of pesticides. If a worker does not breathe, eat, or allow pesticides to be absorbed through the skin, they cannot do harm. Pesticide poisoning can be of two basic types: (1) acute, and (2) chronic. The effects of acute poisoning are seen and experienced rather quickly. The symptoms may be delayed a bit, but they usually happen rapidly. It is usually very obvious as to what has occurred. Chronic poisoning is usually the result of several small doses, the effects of which tend to build over a long period of time. The onset of symptoms, such as irritability and nervousness, are not very dramatic in their appearance. Health may gradually decline over an extended period. Symptoms may be mild, moderate, or severe.

Mild symptoms include general fatigue, headaches, weaknesses, mild nausea, diarrhea, and perhaps loss of appetite. These symptoms may be attributed to many things other than pesticide poisoning. Moderate symptoms are more pronounced and may include stomach cramps, excessive perspiration, lack of muscular control, mental confusion, difficulty in breathing, and extreme weakness. Moderate symptoms are a bit easier to pinpoint than the mild ones. Severe symptoms leave little doubt that an individual has probably been poisoned. High fever, a high breathing rate, vomiting, muscle twitches, convulsions, problems in breathing, and unconsciousness are indicators of acute poisoning.

AIDING VICTIMS

In the event of poisoning, call for medical help immediately. Provide information on which toxic material was being handled. This may be found on the label or the MSDS. Try to protect the victim from further exposure. Move him or her upwind of the pesticide if possible. Rescuers should do what they can to prevent the pesticide from getting on them. If the victim is conscious, assure him/her that help is on the way. If the victim has obvious amounts of pesticide on the skin, flush it off. Remove contaminated clothing immediately and drench the victim with water. Use a detergent or commercial cleanser if available. Dryand cover the victim to preserve body heat and wait for help to arrive.

TIPS FOR DECONTAMINATION

Following are some of the things that may be done to minimize exposure to the harmful effects of pesticides:

- Handle containers carefully so as not to damage them and cause leaks.
- Read and follow instructions found on the label and/or MSDS.

- Keep food and drink away from handling areas.
- Do not smoke in the work area.
- Follow the instructions carefully when mixing and preparing pesticides.
- Keep people away and decontaminate when leaks have occurred — follow instructions.
- Wash hands thoroughly after handling.
- Wash and rinse equipment after use.
- Bathe thoroughly before going on to other work.
- Use care in opening containers.
- Clean protective clothing and personal protective equipment after use.
- Make sure soap, water, and towels are available at the handling site.
- Store leftover pesticides in a locked storage area designated for this purpose.
- Warn people and mark contaminated areas if cleanup is delayed.

CROP SPRAYING AND FERTILIZING

Many of the safety procedures used in handling and applying insecticides also are applicable when working with weed sprays and fertilizers. Chemicals used for these operations should always be handled and applied in accordance with the manufacturer's instructions. Take the time necessary to read and understand all precautions that apply to the chemicals being used. Make sure all involved workers understand what is to be done and how to do it safely. Beginning work without first understanding the nature of the chemical and how to safely apply it can be a very costly mistake. It may be safer and less expensive in the long run to have the crop chemically treated professionally by an organization that specializes in such a service.

Anhydrous ammonia is a chemical that is often applied to croplands in the fall. This chemical, symbol NH_3, is a nutrient essential to plant growth. It is also a hazard to the environment if released in quantity. Even a small exposure to the skin of humans and animals can cause severe burns. A release into the face of a person or breathing its vapors can be fatal. NH_3 is in a liquid form when stored under pressure. When released to the air or soil, it becomes a vapor. Suppliers of this chemical are usually well acquainted with its safe transport and handling. They, as well as the manufacturer's printed instructions and MSDSs, are sources of information on NH_3 safety. This substance is very anhydrous — it actively seeks out water. If it gets loose, it can quickly enter the body through the eyes, lungs, and skin.

Injuries often result from using an improper tank to store NH_3, filling a tank to over 85% of its capacity and using hoses not designed to handle the chemical. Failing to bleed off hose pressure and leaving valves partially open have caused NH_3 to be released.

Because of the hazardous nature of NH_3, many proprietors elect to have this chemical professionally applied, thus transferring a large part of the risk associated with handling and applying it. If this chemical is to be applied by a proprietor and his/her workers, safety becomes a first consideration.

Critical to safe application of NH_3 is to plan and use appropriate work practices. Personal Protective Equipment (PPE) that is designed to protect people from the harmful effects of this chemical should be used. Always have a supply of fresh water on hand when working with this substance. At least 5 gallons of water should be kept in the supply tank and a squeeze bottle of water should be carried on the person(s) handling this material. As an added precaution, a container of water should be kept on the tractor, within easy reach of workers. The best respirator for this chemical is a full-face type which would prevent the substance from being splashed on the face or in the eyes. Lined rubber gloves should be worn along with nonabsorbent boots. Wear a long-sleeved shirt, long trousers, and an apron that is resistant to chemicals.

Hoses, valves, and connections should be inspected for serviceability prior to use. Personnel should stay clear of valve and hose openings. An 85% bleeder is best used when filling the tank. Gauges are often in error. Make sure the hitch to the tractor is a proper one. Safety chains are a must. If the application rig is moved on the public roads, make sure a SMV emblem is displayed. Proprietors should assure themselves that all workers applying NH_3 know what to do in case of emergency. Training in this area should be considered mandatory. Other than the labels and MSDS, the county agricultural agent, the supplier, and often the community college, can be of assistance. Never allow workers to handle or apply NH_3 if they are unsure of what to do in case of emergency.

A breakaway valve is a good safety feature on NH_3 rigs. These valves help prevent uncontrolled releases of the chemical by separating and stopping the flow before excessive force is placed upon the hose. These valves have a limited life span. They should be replaced at least every three years. The ammonia and certain additives that may be a part of the formulation adversely affect rubber and some metal parts. Follow the valve maker's instructions on what to do if the application device becomes loose and the breakaway valve functions incorrectly. Certain pressure bleeding and reconnection procedures must be carefully followed.

If the hitch becomes severed without a breakaway valve, NH_3 will spew out of the severed tank hose. The operator, if alone, should not attempt to correct this situation. Though costly, it is probably better to allow the ammonia to flow from the broken connector hose until the tank is empty. This applies to an unpopulated area. It is usually suicide to attempt to gain control of a ruptured hose without wearing a self-contained breathing apparatus (SCBA) and a protective rubber suit. These items are usually not readily available during the application process. If a rupture occurs in a populated area, call the fire department and allow them to handle the problem. Keep everyone away from the leak. A leak in a populated area is a severe liability exposure. A properly operating breakaway valve should eliminate this type of exposure.

CHEMICAL CHOICES AND AMOUNTS

The type of pesticides, fertilizers, and other agricultural chemicals that were used last year or the year before may not be ideal for the current year. Use professional help available from suppliers, farm literature, and the county agricultural agent to determine the best materials and the quantities needed for the current season. An inquiry may indicate a need for change. Perhaps a different substance will work better, or quality changes may be indicated. If changes in products are made, a careful review of the product safety data should be made and shared with those who will be applying it. Changes might be less expensive, safer, and may do a better job than previous applications.

NOISE AND DUSTS

Unwanted sound is often referred to as noise. Excessive noise can damage the hearing organs and hearing loss is usually permanent. OSHA has specific guidelines that those subject to its regulations must follow. Though a given farm may not fall under OSHA's jurisdiction, their standards may be used as guidelines.

The intensity of sound is measured in decibel (dB) units. Measuring devices can hear sound as low as 1 dB and as high as 140 dB or more; 140 dB is known as the pain threshold, since physical discomfort is felt at or even below this reading. As a comparison, a person standing beside a running jet engine would receive about 130 dB. Sound intensity in a room where people are talking measures about 60 dB. Inside an insulated tractor cab the reading would be about 80 dB when the tractor is pulling a load. Some grain elevators may have a sound level of 95 dB.

Shortly after OSHA became law, it was determined that a person could work for 8 hours at a 90-dB sound level without appreciable damage to his/her hearing. Above that figure (a time-weighted average) either the sound has to be reduced by some means, or the worker has to wear hearing protection. In any case, a hearing conservation program needs to be in place which involves audiometric exams for employees. Today, many organizations require hearing protection if the 8-hour time-weighted average exposure to a worker is 85 dB or more.

Unless sound levels can be reduced by some means, workers exposed to a time-weighted average of 85 dB or more for 8 hours should be wearing hearing protection. Noise exposure may be reduced by both engineering and administrative means, such as putting a muffler on an engine or moving the noise source away from where people work.

Many types of hearing protection are on the market. The most popular are earplugs and muffs. Disposable plugs may be discarded at the end of the day, which eliminates having to clean and sanitize them. In order to promote the faithful use of these devices, allow workers to choose the type of hearing protection they wish to use. Some prefer earmuffs. These, if kept clean, will have a relatively long service life. Muffs keep dusts and pesticides out of the ear a bit more efficiently than plugs, provided the device is properly worn. Muffs have to be cleaned and disinfected daily to remain serviceable. Do not stick cotton in the ears and expect to conserve hearing ability by so doing. High-frequency sounds, which supposedly cause the greatest damage to the hearing organs, go through cotton with little difficulty. Hearing protection allows lower-frequency sound to pass through, such as the human voice. Once people get used to wearing hearing protection, they can converse with one another with little difficulty.

Is hearing protection needed? The only way to find out is to measure sound levels at various places where people work, calculate the 8-hour exposure, and go from there. Of course, many farmworkers are on the job longer than eight hours per day during certain seasons, but they may move back and forth from noisy to less noisy areas several times a day. This should be considered when computing the time-weighted average.

The noise reduction rate (NRR) should be considered when purchasing hearing protection. A piece of hearing protection equipment may reduce noise by 17 dB to as much as 33 dB. The NRR should be used to select equipment after time-weighted averages have been computed.

Dust exposures are nearly always present on farms. These dusts may take the form of mold spores, animal hair, animal waste, and plant particles as well as blowing soil. Lung diseases such as black (coal miner's) lung, asbestosis, and silicosis are all caused by dust particles which have been inhaled and trapped in the lung. Another industrial pneumonia that affects farmers is the result of exposure to grain dusts. This disease is called farmer's lung. Another serious illness that sometimes affects farmworkers is mycotoxicosis. The inhalation of spores that grow in moldy hay cause this disease. These two agricultural lung diseases can cause bronchitis, shortness of breath, and sometimes a lowering of the body's immune system. The latter two lung diseases may be prevented by administrative and engineering controls. Exposure to the causative agents may be greatly reduced by use of respirators and staying out of dusty areas as much as possible. Engineering controls involve such things as dampening feed lots and yards. Not storing wet hay is an administrative control.

Fortunately, the body gets rid of most dust. Nasal hairs catch it, it may be coughed up, it may be expelled when people sneeze, and the like. It appears to be the very tiny dust particles that get deeply into the lungs that cause most of the problem. Particles that are some 5 millimicrons in size, such as the dust one can see floating in the air in direct sunlight, are probably the culprits. Once they are deeply into the lung, coughing will not expel them. Sometimes, the lung tissue walls them off, with the loss of lung capacity as a result. Loss of this capacity is usually permanent. Other times the lung may build fluid around the particles, which presents a pneumonia-like condition. If dust cannot be controlled by administrative or engineering means, the use of respirators may be indicated. It is well to remember that every time dust-producing materials are moved or handled, more dust gets into the air. Prior planning may reduce the need to repeatedly handle such materials.

Normally, a dust filter mask is all that is required for protection in dusty conditions. These masks are easy to wear and many types are available, often in drug stores. The safety supply outlet and the county agricultural agent may be able to advise on which type is best for a given location. The mutual aid group may have a mechanical sampler to measure dust concentrations. Devices are available, although they are expensive and require calibration and care, but it is the only sure way of knowing the concentration of dust in workplace air. Many poroprietors use a rule of thumb: whenever dust is visible in the air, dust masks or respirators are to be used.

No mask or respirator will function properly if it does not fit securely. Beards and facial hair make it difficult, if not impossible, to achieve a proper fit.

COLD WEATHER OPERATIONS

Freezing temperatures may present additional hazards with which farmworkers have to contend. Some of these include:

- Animal water supplies may freeze up.
- Mud may freeze tractor and wagon tires to the ground.
- Icy conditions may present slip and fall hazards to both man and beast.
- Workers may sustain lung damage from breathing freezing air.
- Ice may collect in hoppers and other machinery and could damage equipment.
- Nonwinterized engines may refuse to start.
- Block damage may occur in engines with insufficient antifreeze.
- Frostbite may occur to exposed flesh.

Cold weather is less of a problem in areas where temperatures do not go below freezing. In colder areas, many proprietors miss the straw piles — they rarely exist any more. Straw piles left over from threshing used to provide a good windbreak and shelter for farm animals. The combine effectively did away with the straw pile. Fabricated walls may be used as windbreaks for animals.

The most effective way to deal with the problems of cold weather is to properly prepare for them. The best time to test antifreeze is before the weather gets cold. Covering areas where snow is not desired is hard to do in a blizzard. Good drainage is important to keep ice from forming over large areas. People should be warned about wind chill and how the equivalent temperature may be much lower than that indicated on the thermometer. Many newspapers and periodicals publish a wind-chill chart in the fall of the year. This chart shows equivalent temperatures based upon the actual temperature plus the effects of the wind. When winds are calm, the equivalent temperature and the actual temperature are the same. On the other hand, if the recorded temperature is 10°F and the wind is blowing at 20 mph, the equivalent temperature (ET) is –25°F! At this ET, exposed flesh may freeze in as little as one minute. As wind velocity increases, it lowers the ET. At the measured 10°F temperature used above, a 40 mph wind would lower the ET to –37°F. Wind speeds greater than 40 mph have little additional effect. All workers should be cautioned about working in the cold with exposed skin.

SUMMARY

Environmental concerns are a popular topic today. The Environmental Protection Agency (EPA) and other private and governmental organizations have helped to focus attention on environmental topics. This perhaps happened none too soon. The number of new chemicals and other substances coming into use every month has increased dramatically over the past few decades. Control and cleanup efforts have to keep pace with increased use. Farm people, who live close to the soil, may be some of the first persons affected by toxic agricultural substances. What is done today in this area may have effects upon generations yet to come.

Farm people work in a changing environment and are exposed to all the extremes of it. In addition, farmworkers are often alone while doing their jobs. A lot of the responsibility for environmental concerns rest with this person. Pre-task briefings are very important for farmworkers as their work shifts from one area of responsibility to another.

Environmental safety on the farm contains a liability feature wherein third parties may be involved. For example, pollution of a stream that runs through a neighbor's property may get a proprietor into trouble with governmental authorities as well as a legal action with the neighbor. Cleanups are very costly. It is far better not to get into this kind of trouble in the first place.

If a local mutual aid group exists, a lot of information and lessons learned may be shared. Work planning should be such that it helps to preclude unfortunate incidents that have happened to others from recurring in a given farm operation. The old saying, "There aren't any of us as smart as all of us," quoted earlier, is very appropriate here. Information sharing between farm operators is relatively easy since farmers normally do not directly compete with one another.

FURTHER READING

Atlantic Canada Farm Safety Web Site, Skin Cancer, Another Farm Occupational Hazard, and Safe Pesticide Handling, (*http://www.virtuo.com/farmsafety/health/skin/html*) November, 1998.

Maher, G., Breakaway Valve Will Give You a Break, Agricultural Engineering Extension, North Dakota State University, Grand Forks, News release, March 25, 1997.

Purschwitz, M., Summary: Farm Injury Facts, University of Wisconsin Extension, Madison, 1997.

CHAPTER 7

Livestock Handling Safety

Injury and damage may result from working with livestock. Like many other aspects of farm safety, no single data source is available that details a ranking of accidents and injuries, countrywide, resulting from working with livestock. Indeed, some regional statistics show livestock handling to be the most dangerous work a farmhand does, but most databases show livestock handling in third place, behind tractors and other machinery. In any case, enough data exist to suggest that livestock handling is work that must be done with care to avoid injury, damage to property, and harm to the animals themselves. People who have worked with farm animals for years continue to be amazed at what animals do under differing circumstances. Beef cattle and swine will be the primary animals discussed here.

ANIMAL BEHAVIOR

A basic understanding of animal behavior is necessary to successfully work around them. Most farmworkers who handle animals learned to do so by watching and learning from others. Others may have taken courses in animal management in high school or college.

All livestock tend to have both a maternal and a territorial instinct. Both cows and sows have been known to go to some length to protect their offspring from what seems to them a dangerous situation. If a cow and her calf have to be moved, it is usually more easily done if they can be moved together. Cows become rather defensive and unpredictable when separated from their young. Brood sows have been known to attack humans if they feel their piglets are in danger. When one of her piglets squeals, a sow becomes very agitated and may attempt to bite the worker who is close to the action. Sometimes a farmworker must enter a brood pen with a sow and her piglets. Usually the sow won't mind the intrusion, so long as the worker does not disturb the little pigs. If a worker should step on a piglet and it squeals, he/she may well expect trouble from the mother. Many workers have told of being chased out of brood pens by sows. Cows appear more docile than brood sows, but care must be taken when dealing with animals in the presence of their young.

The territorial instinct in most livestock is strong. They soon develop a routine of coping with their environment and their keepers. Cattle have been known to form a strong attachment to their pens and pastures. A break in movement routine, for example, can be very upsetting to livestock. Animals recognize their herdsmen and are quick to notice strangers in their areas. Young animals can form relationships with handlers as easily as they do with other animals. Unfamiliar places, such as barns, gates, and feedlots are disturbing to animals. Loading chutes can cause problems since there usually is some yelling, more people around, and more noise. These are routine-breakers, and the animals react accordingly.

Workers should be aware that artificial lighting is often a disturbing factor to animals. Sometimes loading animals at night is best. If possible light the area prior to loading, animals are less disturbed than when lighting is rapidly changed during the process.

Both hogs and cattle are generally colorblind and have poor depth perception. Thus they are very sensitive to contrasts. They may shy away from shadows and changes in lighting. Sheep, too, are considered colorblind but they have rather good depth perception. Cattle have a panoramic field of vision. For this reason they can usually see everything around them. Their only blind spot is directly to their rear. The animal is less startled if they are approached from the front or the side. While horses kick to the rear, cattle kick forward as well as to the side. If a cow has a wound or bruise on one side, they are prone to kick on that side. Given a choice, approaching a cow on the side opposite the wound or infection is best. Sometimes animals develop individual patterns of kicking or biting. New handlers should be made aware of these particular animals.

Most animals, including cattle, are very sensitive to loud noises and may spook easily under these conditions. When agitated, animals may move as a group and run over or crush people who are near them. These animals are reacting to fear and may injure handlers in the process. Safety shoes or boots are a good investment for animal handlers. Those who work around animals should be aware of the risks and should do whatever they can to keep from getting into a position where they can be easily injured.

PECKING ORDER

Generally, animals that were raised on the site are easier to handle than those who were added later for feedlot purposes. The raised animals tend to develop stronger bonds with handlers and often take leadership roles within the groups. Groups of animals develop a rather rigid system with respect to one another. People who tend them can observe much of this. In pasture, the same critter will lead the group from place to place for feeding. This same leadership may be observed as cattle make their way to water and feed. Animal leaders tend to discipline others when needed.

Herdsmen identify the animal leaders and form bonds of respect with them. If they can get the leader to do what they want it to do, the others will often follow. These relationships save a lot of time and make the work of the herdspeople easier.

The fact that animals follow a leader may be seen from the paths they wear into the pasture in their single-file travel.

Why this discussion of animal leadership? The wise herdsman will use what he/she knows about a particular group of animals to make the jobs of feeding, spraying, and loading easier and safer. When animals are treated properly, they are much more predictable and easier to handle. Some handlers try to battle a stubborn animal. This often results in injury to both parties.

Animals normally do not try to injure handlers. Occasionally, one that is very provoked will bite or kick a handler, but this is an exception rather than the rule. Animals have been known to crush a handler when they are thirsty and going to the watering device. Sometimes when spooked, they will move as a group and run over a handler.

Animal behavior is often affected by weather changes. In colder climates where snow and blizzards are not uncommon, cattle have been known to break down barricades and fences in an effort to escape cold and snow. Sometimes it's necessary to drive cattle behind windbreaks or on the downwind side of sheltering tree stands during blizzard conditions. Many will not seek shelter without help. Left on their own in such weather conditions cattle will often walk as a group, flattening fences as they go. They sometimes walk for miles. With their backsides to the wind, they have been known to walk over rather forbidding obstacles, injuring themselves in the process.

LIVESTOCK FEEDING AND HANDLING FACILITIES

Facilities built with efficiency and safety in mind can be a major factor in accident prevention. Corals, pens, feedlots, and chutes should be designed to make the animal handling tasks easier and safer. Means of egress for people working with the animals should be considered an important factor for the sake of safety. First of all, workers should do all they can to stay out of harm's way, but failing that, the facility should be designed to provide a means of escape for herdspeople. Man enclosures should be built into places where means of egress are limited. Obviously, more contact between people and animals increases the chance of accidents. Proper work planning can often reduce the contacts that people have with animals, thereby reducing exposure to risk.

Cages or holding stations designed to immobilize an animal while tags are being applied or vaccinations administered should be in good repair and work properly. A drop or two of oil on turnbuckles and latches will help ensure that the work will go smoothly. Injuries to both man and animal have occurred when holding stations malfunctioned. The time to inspect and repair these facilities is before they are needed.

Animal facilities should be illuminated, if needed, during hours of darkness. Workers need sufficient light to work safely. Trip and fall hazards should be removed from access ways.

STRAY ELECTRICAL CURRENTS

Stray voltage may affect animals if a current difference exists between two or more animal contact points, such as a feeding floor and a water cup. When contact is made between these points at the same time it is possible for a small amount of electrical current to pass through the animal's body. Under normal conditions such current flow does not take place. This problem is more common in dairy operations since milking parlors may have several current sources. Stray voltages can result in substantial losses, although it is not a common occurrence.

Reluctance on the part of animals to drink from watering points or to consume feed may be a symptom of stray voltage. Please bear in mind that unusual animal behavior may be due to causes other than stray voltage and a thorough investigation should be made to determine such causes.

Stray voltages may be caused by sources that are within, or outside the control of the proprietor. Causes on the farm may be related to improper grounding, improper wiring, or electrical shorts, to name a few. If two or more circuits are supplying electrical current, unbalanced loads may cause a problem. A qualified electrician is a primary source of help if stray voltages are thought to be a problem. Often, outdoors and in-barn electrical devices are exposed to more severe conditions than in a home. Proper grounding is a very important consideration, not only from a stray voltage standpoint, but also as potential source for electrical shock as well. Proper grounding should be checked periodically as a part of the safety inspection process.

The utility company that provides electrical power to the farm can provide assistance with stray voltage problems, particularly those that originate off-farm.

ZOONOTIC DISEASES

Zoonotic diseases are ailments that can be transmitted between animals and humans. Among the most common of these are rabies, brucellosis, and ringworm. People may get these illnesses by being bitten or coming into contact with infected animal tissue. Ringworm, a fungus, is rather common in cattle. If a person has an open sore or cut, handling of cattle may provide an opportunity to get ringworm. Anthrax may be a problem disease in some areas.

Brucellosis, sometimes called Bangs disease, causes early abortion of the fetus in cattle. When humans get this disease it is often called undulant fever. Drinking milk from an infected animal may transfer the disease to a person. Care should be taken when handling aborted tissue. This disease has been controlled to a large extent, but it is still around. No effective treatment is available for this ailment in animals. It has been controlled in some jurisdictions by state-mandated vaccinations of heifer calves.

Anthrax is a bacterial disease that affects the skin of cattle and sheep. Symptoms in humans are mild at first, often like the common cold. This may be followed by acute symptoms such as respiratory distress, shock, and fever. This disease may cause death. Contact with animal hair, hides, bone products, and wool exposes people to this disease. It is not transmitted from person to person. Preventive vaccinations are available. Other preventive measures include dust control, limited handling of hair, wool, and hides, plus good personal hygiene.

Leptospirosis and salmonellosis are other zoonotic diseases that may be passed from animals to humans. Leptospirosis may manifest itself in several forms including fever, headache, chills, severe malaise, vomiting, renal insufficiency, hemorrhage, and occasionally jaundice. Rats and pets, as well as farm animals, may carry this disease. It is transmitted through contact with abraded skin or contact with mucus membranes in water. Transfer from person to person is rare. It may be prevented by protective clothing, boots, and rodent control.

Salmonellosis is a bacterial disease that promotes acute enterocolitis, headache, abdominal distress, diarrhea, nausea, and vomiting. Septicemia (blood poisoning) may be a manifestation. Certain types of *Salmonella* that infect people and animals cause it. Domestic and wild animals may carry the disease. Eating food contaminated by feces of infected animals may transmit it. It may be spread by drinking contaminated water. Proper handling and cooking of foods will assist in controlling this disease.

Zoonotic diseases are still around, in spite of efforts to control them. Donna Sue Dolle, M.D., P.A., who is board certified in internal medicine, reported involvement with several cases of these types of diseases while serving as chief resident at the University of Texas Medical Branch at Galveston.

Personal hygiene is a first line defense for avoiding these diseases.

OTHER ANIMAL AILMENTS

Feedlot lameness involves the animal's feet. Foot diseases account for most lameness in feedlots. Most are caused by sole punctures, excessive wear, toe abscesses, or injury to the foot. No program of treatment should be started without a veterinarian's diagnosis.

Toe abscesses frequently occur when young cattle from good pastures are placed in feedlots. Penetration of the sole may lead to infections under the hoof wall. Usually these abscesses occur first at the outside rear toes of the animal. Infections of the outside front toes are usually most severe. Erosion of the sole may be the result of having a too clean feedlot. Dirt and dried manure act as a cushion for the feet of these young animals, who are used to walking on soft earth. Cleaning and rains may wash the cushion away.

Early symptoms are careful walking and short steps. The feet do not swell in the early stages. Almost all animals treated before more pronounced symptoms appear to recover rather quickly. It is definitely best to treat affected animals early.

Toe injuries may be prevented by eliminating toe traps. Spaces between the ground and walls may provide a suitable space for an animal to injure itself. Mechanical injuries caused by stepping on trash or metal should be promptly treated, and the causes eliminated. A veterinarian can advise proper treatment.

Foot rot develops following an injury to the soft tissue between the animal's toes. This ailment is easily misdiagnosed. It has been estimated that only a small percentage of lameness is caused by foot rot. Stubble or frozen mud often causes injury to the soft tissue. As with many ailments, the veterinarian is the best source of diagnosis, information, and treatment options.

Fortunately, drugs are available that can successfully treat a host of animal ailments. Bear in mind, however, that drug use is not a substitute for good livestock management.

SUMMARY

Animal safety on the farm is closely tied to human safety. While protection of people from harm is a governing concern, the safety and well-being of farm animals is also an important consideration. The costs of producing a pound of beef or pork are high. Animal diseases and accidental injury add measurably to these costs. Although some salvage value is received for a killed or injured animal, it is not as much as that received for a healthy animal.

Proper handling of livestock requires skill and experience. Good herdspeople produce profitable herds. They respect the animals, and the animals respect them. They keep themselves and their animals out of harm's way.

Preparation and/or repair of feedlots are best done prior to the start of the feeding season. Having to work in close proximity to animals adds risk to the job. Proper planning and scheduling can keep people/animal contacts, and therefore accidents, to a minimum.

Those who are new to livestock handling would do well to spend some time with an experienced herdsperson. Those who work with animals need some understanding of animal behavior.

With today's more crowded facilities, dusts generated by feeding operations and dried waste may result in respiratory problems for animal handlers. Personal protective equipment such as dust masks and eye protection should be provided and worn when conditions merit. Alertness and personal hygiene go a long way in the prevention of zoonoses — animal diseases that may be transmitted to humans. Some of these illnesses are severe, such as anthrax, rabies, and brucellosis. Some zoonotic diseases are spread by inhalation of dusts from contaminated manure or soil.

Always check with a veterinarian for diagnosis and treatment of animal ailments. Sooner is usually better in this regard.

FURTHER READING

Bean, T. L., Working Safely with Livestock, Factsheet AEX 990, Food, Agricultural and Biological Engineering, Ohio State University, Columbus, 1990.

CHAPTER 8

Materials Handling

Improper material handling may result in injury or property damage. Agricultural operations employ many types of material transfer. One of the simplest ways to avoid accidents in this area is to cut down on material handling through work planning and anticipation of need. For example, if seed corn can be moved on to the site and then moved only once more to the planting area, a lot of unnecessary handling could be avoided. Without prior planning, seed bags may be dropped off in the yard by the supplier, moved to temporary storage, moved again to a more permanent storage, and finally moved to the planting areas. Each time material is moved, risk of accident is present. Planning not only addresses safety concerns, but also reduces labor. The following apsects of material handling will be discussed here:

- Manual handling
- Mechanical handling
- Ergonomic considerations

MANUAL HANDLING

Back injury relative to manual handling operations is always a concern. Back injuries are often the result of improper lifting and failure to understand and appreciate the mechanics of the back. In industry, back injury claims are many and complaints about sore backs are common. Perhaps thousands of workdays are lost each year because of back injuries and back injury complaints. Many industrial groups recognized this problem years ago, and steps were taken to help reduce this type of loss. Many training programs were developed and several lifting techniques were advocated. These frequently produced unsatisfactory results. Often the worker was bombarded with a lot of facts, figures, and geometric jargon that he/she did not understand. In many cases the worker couldn't lift in the manner the new systems specified. Some would fall over trying to get into the lifting position. It is thought that most workers threw up their hands and went back to lifting the best way they could. They continued to experience back problems.

Proper lifting is not difficult. When many studies and much experience are boiled down to their simplest terms, the following tips will greatly assist in reducing injury to the back during lifting operations:

- Lift with the leg muscles rather than those in the back.
- Lift no more weight than you can safely handle.
- If you must turn with a picked up load, do so with the legs and feet.

The human back was not designed to be a lifting device. Viewed from an engineering standpoint, the back is a horrible design for lifting tasks. When a person uses back muscles to lift, he/she is using those near the base of the spine. In this instance, the spine may be viewed as a lever with a short power arm and a long resistance arm. Hardly anyone would be foolish enough to use a pry bar over a fulcrum with a 6-inch power arm and a 4-foot resistance arm. It wouldn't work. Yet, this is what happens when a person tries to raise a load by using the back to do the lifting. The power/resistance ratio when lifting with the back is about 1:9. The mathematics alone should prove that lifting with the back makes little sense!

Workers should be shown how and encouraged to lift with their legs. The muscles in the legs are the strongest in the body. A worker should:

- Size up the load.
- Be aware of where the load is to be placed.
- Get as close to the load as possible.
- Take a firm grip.
- Hold the back firmly in its natural position.
- Raise the load by straightening out the legs.
- If unable to raise the load, get some help.

Prior to getting into position to lift, the worker should observe the load from a couple of different angles, as well as survey the route to be taken with it. Is the way clear? It is always better to remove obstructions rather than try to go over or around them. Note likely places to take hold of the load. Gloves may help in securing a firm grip. Some lifting instructions tell the worker to "keep his/her back straight." As such, this is an impossible task. "Hold the back in its natural position" is perhaps a better instruction. Once a firm grip is attained, the lift should be attempted by using the strong leg muscles. If it becomes obvious to the worker that he/she is unable to safely lift the load, leave it and try to get some help. Getting help is not easily done on the farm. Oftentimes, workers are alone on a job. They should be instructed to take no unnecessary chances with their backs and to go find some help.

Proper lifting calls upon the worker to make some judgments as to his/her capacity to lift things. Proprietors would do well to insist that workers make these judgments based upon their own work experiences. How much weight can a human lift? There is no single answer to this question. Supposedly, men can safely lift more weight than women because they are usually larger in stature. Some local studies of this question of weight exist, but none are known to be fully accurate. There are many individual differences. How much weight a given person should be expected to lift is not easy to determine.

Perhaps one of the most stressful behaviors with regard to the back is lifting and twisting, or lifting followed by twisting. This places a great deal of strain on the back. Rather than twisting the back under load, it is much safer to instruct the worker to shift the load with his/her legs and feet.

Proper warm-up is important. Instruct workers to handle smaller loads than normal when beginning a material-moving task. A tip in this regard may be taken from professional weight lifters. These people lift huge weights. Rarely does one hear of a weight lifter with a bad back.

Dollies, hand trucks, and other devices may well save having to do a lot of lifting and carrying. A few dollars spent for such devices can go a long way in easing the lifting and moving tasks. Their use will make the jobs easier and more work can be done in less time.

MECHANICAL HANDLING DEVICES

Forklifts, front-end loaders, power scoops, and self-loading vehicles are referenced here. Equipment such as the above have not eliminated human handling of materials, but such handling has been reduced. With the introduction of this equipment have come additional hazards with which a farm proprietor and his/her people need to contend. One of the primary hazards is people working in close proximity to powered equipment. Because of the proximity, such equipment should be fitted with alarms that would warn workers that the machine is about to back up. Employees should stay well out of the way when powered equipment is being moved from one part of the job site to another.

When front-end loaders are stored or are not in use, the bucket or other devices should be lowered to resting blocks on the ground. This would preclude the raised device (Figure 8.1) from being accidentally lowered onto a person standing beneath it. Normally, the motor need not be running and a child or someone else may accidentally lower the device on an unsuspecting bystander.

Overhead clearances need to be carefully checked. Raising a loader into overhead electrical wires is not uncommon. If such a loader becomes energized, the operator is normally not affected, but if someone should touch any part of the vehicle, he or she completes a grounded circuit and electrocution is a possible result. If the operator brings the machine into contact with electrical wires, he/she should carefully back away and disengage. If any doubt exists as to the success of this maneuver, the operator should call for help. Operators should be reminded of a lesson they probably learned in kindergarten, if you are trapped in a vehicle that has become energized, and if you have to leave that vehicle, jump free. By so doing, the circuit will not be closed and there should be no ill effects.

Grain elevators are material handling devices that may come into contact with overhead electric lines. Movement of such devices and their erection should be planned so that they are routed away from such overhead obstructions. Because of their length, grain elevators have a large tail-swing. Those moving an elevator should at all times be aware of where the discharge end is in relation to people and structures nearby.

Figure 8.1   An improperly stored front-end loader. The bucket and clamp should be at ground
level when not in use.

If the grain elevators are driven by a belted tractor through a power device, make
sure the device is securely staked to the ground. This will preclude its coming loose
when the elevator is under load.

Hay bail loaders are often employed as mechanical lifting devices. Many types
of these exist and all require practice and skill for safe use. Usually, someone is on
the ground near bales being lifted. The operator and this person should maintain eye
contact as much as possible. The person on the ground should not get into a position
where the operator cannot see him/her. Special bail hauler attachments are placed
upon front-end loaders for moving large round bails. This is usually a one-man/one-
machine type of operation. These bails should be carried as low as possible for
reasons of rig stability. In fact, all loads carried on front-end equipment should be
carried as low as is practical. Steering may be impaired, particularly during wet soil
conditions. Operators should move slowly and steadily during these conditions and
be prepared to use the rear-wheel brakes to assist in steering.

When animal self-feeding wagons/hoppers are used, the operator should make
sure all guards are in place and to watch for frozen auger hoppers during freezing
conditions. Augers should be carefully engaged if freezing is suspected. Engagement
of a frozen auger under throttle could destroy expensive equipment.

For the same reasons, silo unloaders should be carefully checked during
extremely cold weather.

MOVEMENT OF PORTABLE GENERATORS

Portable generators not mounted on wheels are often skidded to where they are needed. In colder weather, with frozen ground, it is best to load these on a wheeled vehicle for moving. A front-end loader can be used to pick up a generator. Make sure the pick-up chain or device is hooked to a proper lift point on the generator frame. Do not forget to ground the generator prior to starting it. This can be difficult on frozen ground, but the extra effort is worth it for safety reasons.

ERGONOMICS CONSIDERATIONS

Ergonomics pertains to the many interactions of people with the total work environment. Heat, cold, dusty conditions, height of workstations, and location of switches and controls are only a few of the items considered in ergonomic studies. Ergonomics is a British term. Human factors engineering is the term most used in the U.S., although the British term is gaining in common use. Biomechanical engineering is the term many prefer and is perhaps the most descriptive of what ergonomics is and does.

Why is ergonomics a concern? It grew out of an assessment of a basic fact of life: *one may modify the workplace at will, but human beings cannot be reengineered.* Rather than locating a machine control where an operator has to stoop to use it, place it where it is convenient to the user. Though not as true today as it once was, controls on machines sometimes appear to be added and placed as an afterthought. When a designer plans a new production machine, he or she can sentence an operator to a lifetime of aches and pains because of the poor location of controls. Since human operators cannot be given longer arms or shorter legs, designers must modify the machine to accommodate the people who are going to operate it.

It is unfair to place poor design of the workplace solely upon the heads of designers and engineers. Proprietors have a role to play and workers do as well. Many modifications may be made in the workplace to accommodate physical differences between workers. For example, a sturdy box might be used for a worker to stand upon if he/she is too short to comfortably operate a piece of equipment. Workbenches may be modified to allow them to be raised or lowered to accommodate people. Nonskid surfacing materials may be applied to combine ladders to help preclude slips and falls.

Ergonomic principles may be applied by anyone who is trying to make the workplace and its tools better fit the worker. Workers should be encouraged to submit suggestions for improved workplace design. Those who do the work soon learn that ergonomics makes their jobs easier and often more pleasant. Obviously, there are economic restrictions on making lots of workplace changes. When new equipment is contemplated, the proprietor would do well to take ergonomic considerations into account as a part of the decision-making process.

SUMMARY

Material handling is an important part of all agricultural operations. One of the simplest ways to reduce material handling accidents is to cut down on the number of instances where it is required. This may be accomplished through work planning and scheduling. If a certain material can be handled twice rather than three times, exposure to accidents is reduced by one third, assuming equal hazards each time. When materials are brought on to the site they should be located in a position where they will not have to be moved again prior to use.

Prior planning is sometimes dismissed as too time-consuming, or not worth the effort. This is simply not true. Usually a multiplier effect occurs from planning a job to time spent on that job. If 10 minutes are spent in planning, it is not unusual to save at least three times that much time when completing the task.

Labor-saving devices have speeded up work and saved people from exposure to back injury. Although mechanical material handling equipment may bring hazards of their own to the workplace, they have a net positive effect on reduction of injuries to the human back. Ergonomic studies point out that the workplace may be modified to better fit the worker, but it is impossible to modify man. People are essentially creatures of habit. This is often viewed as resistance to change. Changes in order to make things better should be initiated as the need arises. Nobody ever solved a problem by trying harder to do what doesn't work. If a material handling operation is presenting problems, change it. Use worker input in order to accomplish this.

Even with the best mechanical engineering it is perhaps impossible to eliminate all need for human lifting and handling. Proper lifting techniques discussed in this chapter need constant attention by those who engage in them. In organizations where people do a lot of manual handling of materials, workers often coach one another as to proper lifting techniques. In the midst of a busy day, workers may forget to lift in the proper manner. They normally appreciate reminders. As in many safety issues, the boss should set the example. People learn from watching others, as well as from instruction.

FURTHER READING

Barenklau, K. E., *Measurement of Ergonomic Safety Programs,* Proc. 2nd Annu. Int. Symp. Ergonomics, American Society of Safety Engineers, Des Plaines, IL, 1994.
MacLeod, D., *The Ergonomics Edge*, Van Nostrand Reinhold, New York, 1995.

Activities Designed to Assist the Farm Proprietor in Preventing Accidents by Promoting Work Safety

CHAPTER 9

Safety Communications

There are probably nearly as many definitions of communications as there are groups discussing the subject. Sometimes communications are used as excuses when something goes wrong. The person who made the error, or failed to communicate, can look around rather hopelessly and say, "communications." Perhaps it is easier to blame communications than to take responsibility for action. For purposes of discussion here, the following definition of communication will be used:

Communication is what we do to give and get understanding.

—Frank E. Bird, Jr., P.E.

Although language is the usual form of communication, other methods such as media, signals, and body language, to name a few, also may be used. An example of the use of media in safety communications is the safety warning sign and the poster.

POSTERS AND SIGNS

Do posters do any good as far as safety is concerned? Yes, when they meet a couple of criteria. In order to be effective, a poster must do the following:

- It must define a problem (or opportunity).
- It must tell people what to do about the problem/opportunity.

Farming operations may not use as many posters as a factory, but they can help communicate a safety message that workers and others may need. Posters serve as timely reminders and may show how to solve a given safety problem or concern. They are available from many sources. The local safety equipment supplier may have them, and if not should be able to recommend a source. The National Safety Council, Chicago, has offices in many states. They are a source for safety media of

81

all types, including posters. Posters may be left up for 90 days before they start losing their effectiveness. Any time they get dirty, wet, or wrinkled, they should be taken down.

Other forms of safety media are signs that provide information or a warning. A sign such as, "Wear eye protection while grinding" might be placed near the bench grinder in the shop. The brightly colored signs that mark the location of fire extinguishers are another example of a helpful media device. Do not listen to the old argument that everybody knows that anyway. Even the most safety conscious people need reminders from time to time. Posters and signs are a rather inexpensive way to communicate safety information. Children who live on farms often make safety signs and posters at school for their parents.

Safety newsletters are good communication devices. While it may be impractical for an individual farm to publish its own, if a mutual aid group exists a newsletter is a worthwhile project for them. If all farms in the group submitted safety items, such a paper would enable all to learn from the experiences of a few. If a mutual aid group does not exist in a given neighborhood, it would be a worthwhile effort to start one.

INSTRUCTIONAL COMMUNICATIONS

People need to be taught how to safely and effectively accomplish tasks. It has been said that experience is the best teacher. This may or may not be true. Experience is probably the most expensive teacher. Left to his/her own devices, a worker will eventually learn how to properly do a job. He or she may, at the same time, leave a trail of destruction behind while learning the correct way to do a job on his/her own. Three topics will be considered in this section on instructional communications:

- Effective job instruction
- Safety briefings
- Safety reminders

Much of what has been learned about teaching people to do things while on the job occurred during the rapid industrial buildup during the early days of World War II. This rapid industrialization became a success story that has hardly been surpassed. One of the primary reasons for its success was the development of an on-the-job training system. Most of the U.S. population lived in rural communities at the time. How does a man or woman who has never done anything more technically difficult than cinching up a saddle on a horse get to be an accomplished welder on an aircraft fuselage? And do it within a couple of weeks! After only a few fits and starts, adult educators developed the on-the-job training system that worked well and is still used today.

How does one go about teaching people to do a job they have never done before? One of the easiest ways is to use on-the-job training techniques. The first step in teaching is to get ready. Here is a way to do that:

- Make a teaching plan.
- Break the task down into consecutive steps.
- Have all the materials needed available.
- Select a site and arrange it.

What do we mean by a teaching plan? A lesson plan such as a professor would use? Of course not! In fact, this plan may not even need to be written down. What is important is how will the person doing the instructing get another person to do the task correctly, safely, and quickly the very first time he/she does it. Obviously, the teacher has to know the job very well. A person who does not know how to do the job should never try to teach someone else. This sometimes happens, regrettably. Part of making the plan is learning the job. What is a job/task? For purposes here, a job/task is usually a seven- to eleven-step process that either completes the work or moves it along. Carpentry is not a job. Masonry is not a job. These are occupations. Hanging a door, however, is a job or task that a carpenter may do. The preparation steps are very important in getting ready to teach a job. Getting it done right the first time should be the goal of the instructor.

EFFECTIVE JOB INSTRUCTION

The steps in conducting on-the-job training are

- Tell
- Show
- Test
- Check

On-the-job training normally implies one instructor teaching one person at a time. This is appropriate for farm situations where there are few workers and each of them does many jobs. Safety instruction should be a part of job instruction.

Tell means to instruct clearly and completely, going through each of the job steps. It is well to use the actual piece of equipment involved. People learn through their senses. The good instructor tries to involve as many of the learner's senses as possible. If the instructor just talks, the learner can only use one sense — hearing. If the teacher uses a piece of equipment or a training aid, the learner uses two senses — hearing plus seeing. If the instructor involves the worker in handling the equipment being discussed, an additional sense, that of touch, is used. The other senses, taste and smell, are not utilized as often as the other three, but they may be employed in certain subjects. It is generally better if the teacher goes through all of the job steps during the initial presentation. People tend to learn better and faster when this is done.

The **Show** step may involve a demonstration using the equipment that is the subject of the training. The instructor should move through the job steps, showing the learner exactly what he/she is being trained to do. As an example, following are the job steps for jump-starting a vehicle with a dead battery. Before going through the steps the learner is warned to wear eye protection while doing this job and to make sure that both the live and the dead battery are the same voltage. The job steps are

1. Make sure ignition switches of both vehicles are in the off position.
2. Connect the + (positive) terminal of the live battery to that of the dead battery.
3. Connect the – (negative) terminal of the live battery to a grounded body part of the dead vehicle.
4. Turn on the ignition switch and start the live vehicle before turning on the ignition switch of the dead vehicle.
5. Attempt to start the dead vehicle.
6. When disconnecting, do it in reverse order.

Make it clear to the learner that improperly handled batteries can be dangerous. There should be no open flame around batteries.

Test the learner by having him/her do the job as taught, going through each of the job steps in order. Ask the learner to tell the instructor what he/she is doing as if he/she were teaching the instructor. When this is done, learning is greatly enhanced. People tend to remember over 90% of what they say as they do a task. The military made good use of this method of instruction. Thousands of Army Air Corps pilots were taught to fly using this technique. There is another advantage to the instructor here — if the student makes an error, he/she both does and says it. Errors should be corrected and the learner asked to continue. The instructor should not move on with the instruction until he/she is assured that the learner knows how to do the job. When the learner has demonstrated proficiency, the instructor moves on to the last phase of the training.

Check means to watch the person do the job a few times. If there is any doubt about the learner's ability to do the job safely and correctly, ask him/her to go through it again for the instructor. This should not be viewed as harassment. Most people like to demonstrate their know-how. As the person becomes proficient, the checks may be tapered off. The old admonition, "Be a supervisor, not a snoopervisor," probably applies here.

It is important to remember that safety instruction should be included in job instruction. Sometimes instructors will teach the job and then, at the end, go back and tell the person the safety items relating to it. Do not separate safety from production. Teach people how to safely do a job.

On-the-job training is a necessary activity that is not difficult to accomplish. Reviewing the steps presented above will help those who act as instructors to become more proficient.

SAFETY BRIEFINGS

Safety briefings are short workplace communication devices that are sometimes called toolbox or tailgate meetings. These little information sharing sessions cover one topic and are usually no more than 5 minutes in length. They may be used anytime there is a need to pass on some safety information that will assist workers to more safely and effectively accomplish a task. They are rarely scheduled, but take place whenever a supervisor or lead worker feels that a short get-together will improve safety and efficiency. These briefings need not be limited to safety. They can address production and quality efforts as well. Even though these briefings are informal, a few points, when practiced, will help workers in their tasks. These pointers are

- Prepare for the briefing.
- Zero in on one main point or idea.
- Relate it to those being taught.
- Use equipment or a training aid to enhance the session.
- Drive home the main point.

Good preparation for any sort of safety communication is a must. Seasoned industrial trainers take time to prepare what they are going to present. It takes very little time to prepare a 5-minute safety briefing, and the effort will pay big dividends for both the instructor and the learners. While on-the-job training is often a one-on-one activity, safety briefings work equally well with groups. If a few people need to know a certain point, it is more economical in terms of time and effort to conduct a briefing for all at the same time. If group sessions are not practical, a safety briefing can be utilized as a one-on-one device. Preparation should begin as soon as the person responsible for the briefing feels that a little group meeting is needed.

How does one prepare for a briefing? The following activities may help a person prepare:

- Think about the topic. Bring your own ideas and observations into play. Think about why the workers need the information you are preparing.
- Write down ideas as they present themselves. Carry a small notebook and jot down things that will help people do a better and a safer job.
- Read selfishly, with workers in mind. Clip articles. Organize the information in a form that will allow its use in both job instruction and safety briefings.
- Note what workers and others say about the topic. What has been their experience in the past? Ask workers for their ideas on a specific problem or topic you are wrestling with.
- Outline your briefing. This may be written down as notes to use, or it can be committed to memory. Remember that meetings take time and time is usually in short supply. Lining up your thoughts before making a presentation will save time, and probably greatly assist in doing a good job.
- Practice your briefing. Practice may not make a meeting perfect, but it will greatly enhance the quality. Remember that you are probably not briefing strangers. You know these people and this knowledge can help you do a better job.

Following is an experience that one person had in conducting a safety briefing. The proprietor's son, Jeff, returned a test meter that he had borrowed from another farm. While there, an accident occurred that involved a deep cut on a worker's forearm. The person who was injured dropped to his knees and appeared to be assessing the cut. Two others near the injured person just stood there — sort of stunned — and just watched. Jeff started toward the injured man just as the other two workers came out of their trance and immediately went to the person's aid. They acted as if they were not sure what to do, but finally they got a compress on the cut and someone called the paramedics.

On the way home, Jeff decided that he was going to do a safety briefing for his personnel in case such a thing should happen to one of them. He followed the guidelines above. He began mentally preparing by thinking about how to control bleeding. He decided he would conduct a short briefing on how to control bleeding by the direct-pressure method. He planned to use a demonstrator — someone to play the part of an individual with a bad cut on his arm. He simulated a cut using lipstick. Jeff used a handkerchief as the compress and another to fasten it to the simulated wound site.

He began the meeting by telling the attendees what had happened at the other farm. He involved the attendees by telling them that it could happen here, and all should be prepared to control bleeding. He used the other pointers and drove home his point by saying, "Remember, if anyone gets a bad cut around here, get direct pressure on it immediately and have someone call for help." The briefing lasted no more than 5 minutes and everyone went back to work knowing they had been given some information that may save their lives someday. Everyone appreciated Jeff's effort.

Anyone may conduct a safety briefing. In other industries, workers are often called upon to present such talks. Such talks are often scheduled in advance in larger organizations. On the farm, they are most valuable when used to address a problem or opportunity shortly after it comes up. Safety briefings are a means to promote workplace safety and to give people an opportunity to take care of one another.

SAFETY REMINDERS

An informal, yet effective communications device is the reminder for safety. It is rather normal for people to give one another safety tips. This is done all the time between parents and children. "Wait until the light is green before you cross the street," and, "Don't touch that stove; it's hot and you might get burned," are examples of safety reminders. These reminders are also valuable for adults. A safety reminder is a verbal tip or hint that will help a person to better and more safely do a job.

When is the best time to give safety reminders? Such reminders may be given at any time that someone feels they are needed. Some of these instances include:

- When someone is doing an unfamiliar job
- When a person seems unsure of what to do next
- When there are unusual job circumstances
- When there is an unusual hazard

Everyone may give safety reminders. Almost everyone has had experiences that would help fellow workers if shared with them. Since this is an informal device, no records are kept and no scores are maintained. It is an important activity nonetheless. Reminders contain two parts — the reminder itself, and often an expression of personal concern for the person being reminded. "John, tie that ladder off at the top. I don't want you to fall and hurt yourself," is an example of a reminder. The expression of concern is a nice touch. It avoids the appearance of preaching. Hardly anyone can object to someone else showing concern for him/her. A person may give reminders several times a day if need be. When reminders are sincerely given, this activity often has a snowball effect. Others start doing it as well. When people begin watching out for one another by the use of tips and reminders, the safety effort is enhanced.

LISTENING — A NECESSARY COMMUNICATION SKILL

It is a curious fact that listening is taught as a language art in many public schools throughout the world, except in the U.S. Most Americans spend some 45% of their communication time listening to something, yet most have never been trained to do so. To effectively communicate in safety and in other areas, people need to become better listeners.

Professional counselors, to whom listening is a primary activity, are the experts in this area. They have trained themselves to listen not only to what people say, but also what they mean when they say it. Perhaps all people can do this to some degree, but not to the extent that they would like. People become better listeners through practice and making an effort to improve. Good listeners are patient. They do not interrupt people, nor do they show displeasure while listening. Concentrate on what people say and listen for the why and wherefore. This will not only pay dividends in safety, but in production activities as well.

SUMMARY

Good communication is as important in safety as it is in productivity. Communicating safety concerns to people allows them to share the mission of the organization. Human beings are complex organisms. Misunderstandings are often commonplace because of this complexity. People tend to communicate with emotion as well as logic. This is demonstrated by the old "two people — six messages" saying. When two people are talking, there could be as many as six messages involved:

- What one means to say;
- What was actually said;
- What the other person heard;
- What the other person thought he heard;
- What the other person says;
- What the first person thought was said.

Intonation, speed of speech, and facial expressions can greatly alter the meaning of a simple message.

In this chapter, concentration has been upon a few practical things that may help people do their work safer and better. Good communication techniques are needed when helping a person to learn a new job, or when passing on information that a worker needs to know. On-the-job training is an old tool, but it still does a good job in teaching workers. Safety briefings and safety reminders are practical communication techniques that get needed messages to those who might benefit by them.

People need to know what is going on and how what they are doing fits in with everything else. The wise proprietor will do all he/she can to see that all necessary safety information reaches the workforce. One can hardly go wrong by keeping people informed of happenings at the work site that may affect them.

FURTHER READING

Bird, F. E., Jr., *Management Guide to Loss Control*, Institute Press, Loganville, GA, 1980, chap. 8.

Bird, F. E., Jr. and Germain, G. L., *Practical Loss Control Leadership*, International Loss Control Institute, Loganville, GA, 1992, chaps. 9 and 10.

CHAPTER 10

Safety Inspections

Inspections are one of the oldest management tools. This work activity has been successfully used in production, quality, safety, and cost control efforts in all types of business and industry. In safety, inspections are used to alert people to substandard behavior and conditions that could lead to accidental loss. Inspections are a before the loss activity which, when properly used, alert proprietors to loss potentials.

Several types of inspections are used, some of which will be discussed in this chapter. In order to be effective, inspections have to be properly planned. Even though farmworkers are expected to be on the lookout for substandard conditions at all times, planned inspections provide a systematized search for workplace hazards. Farm proprietors utilize both informal and formal inspections. Who should conduct inspections within a farming operation? The proprietor may perform this function, or may delegate it to someone else. If delegated, the proprietor should make his/her wishes known as to the type and frequency of inspections.

INFORMAL INSPECTIONS

Informal inspections are utilized the most. These inspections are conducted routinely by a person as he/she goes about normal work routines. Those charged with inspections should carry a durable pocket notebook to write down findings that cannot be handled at the time of discovery. Findings of workplace inspections are too important to trust to memory. Notes taken at the time of inspection may serve to jog the memory when findings are reported to the proprietor or to those who will be involved in correcting what was discovered.

Some proprietors concentrate on items to be checked one day each week. Although the proprietor or designated inspector is constantly watching for things which may need correcting, some check certain groups of things on a given day of the week. For example, Mondays may be the day to check for building and ground hazards, Tuesdays for hand tools and equipment, and so forth.

When a hazard is found, and if the person who found it has the resources to correct it, it should be abated at that time. No need to write it down and discuss it;

just correct it. The person who found the hazard may want to inform the proprietor as to what was found and what was done about it. If this hazard has been discovered a few times before, the proprietor may wish to find the reason and take corrective action. If nothing is done about it, the hazard will probably crop up again and may lead to an accident.

FORMAL INSPECTIONS

Formal inspections are undertaken for specific reasons and are usually conducted prior to beginning a major activity, such as planting, harvesting, haying, and receiving feeder cattle. An inspection will assist the proprietor in pointing out things that need attention prior to the start of a major effort. Inspections provide information useful in production activities as well as safety. These inspections may be accomplished as team efforts. In a team inspection, the interaction of the team members and the sharing of information often leads to better and more practical solutions to problems.

As is the case with informal inspections, hazards should be corrected as soon as they are found if resources exist to do so. This saves time and shortens the follow-up list. Those hazards that cannot be corrected when they are found should be noted and responsibility assigned for correcting them. The inspection notes can serve as a guide for correction. By taking inspection notes, the possibility of omitting or forgetting something that needs attention is decreased. The primary purpose of a safety inspection is to find, evaluate, and abate hazards — not to make lists of things that are wrong. Written notes are only useful when they provide information that will aid in corrective activities. Once hazards are corrected, the list may be discarded. In some cases proprietors keep inspection reports in a file for a year or so, and then discard them. This provides some historical data should it be needed later.

HAZARD CLASSIFICATION

A simple hazard classification system will aid the proprietor in setting priorities for correction. Normally hazards are corrected on a worst–first basis. At the start of a busy season, the proprietor could well use a tool that will allow him/her to assess the importance of each item to be corrected. In aerospace, where hazard classification is used throughout, very complex systems are in place. Use of such a sophisticated system on a farm would probably prove to be unwieldy. Following is a simpler system that might be used. Hazards, as discovered, will be classified either "A," "B," or "C."

- A — this hazard is the most serious. It could cause loss of life, limb, equipment, livestock, and/or buildings.
- B — this is a serious hazard which might result in disabling injury or disruptive equipment and facility problems, but it is less sever than "A."
- C — this hazard, if uncorrected, could cause nondisabling injury and/or nondisruptive damage.

Examples of an "A" hazard might be a missing power takeoff guard or a worker entering an oxygen-limiting silo. "B" hazards may include a broken ladder rung or a missing table saw guard. "C" might be ice buildup on a feedlot walkway or a worker handling scrap metal without gloves. All hazards should be corrected, but it is reasonable that, given a choice, the more serious ones should receive priority for correction. A hazard classification system helps establish such priority.

DECISION-MAKING IN HAZARD ABATEMENT

Other considerations besides hazard classification may be taken into consideration by the proprietor when making decisions with regard to abating a given hazard:

- What is the probability that this hazard may produce a loss? High, Moderate, or Low.
- What will be the cost of correction? High, Moderate, or Low.
- If the money is spent, what degree of control will be achieved? High, Moderate, or Low.

If a hazard has a low probability of producing a loss, has a high cost of correction, and the degree of control would be low, this hazard may not be abated as soon as one with a high loss potential and a low cost of correction. If abatement of the latter type of hazard has to be postponed, everyone should be warned about the hazard and other steps taken to prevent loss until it is fixed. All hazards should be abated, but proprietors have to apply their own best judgment when allocating operating funds.

INSPECTION TIPS

Most proprietors and permanent workers are very familiar with the tools, equipment, and layout of the enterprise. Because of this familiarity, it is easy to go through the motions while inspecting and not spot things that should be noticed. Inspections should be methodical and thorough. Look on top of and under things. Open doors and drawers. Look behind things. Look up and around, as well as down. One of the benefits of inspections is that lost tools will be found or misplaced repair parts will be recovered. These are bonus effects of inspections.

Before starting any inspection, two things must be known: (1) what to look at, and (2) what to look for. These two things are especially important in vital part inspections. A vital part of a machine or operation is one that is so critical that if something happens to it a loss will occur. An example might be the fuel injection pump on a diesel engine. Normally, oil changes are performed on these pumps regularly. If the drain petcock is not tightly closed, lubricating oil could leak out and destroy the pump. It rarely happens, but once in a great while a worker may fail to close the petcock after draining the oil, or forget to add oil after draining. A preoperation check (vital part inspection) on diesel engines should include checking to see that the drain cock on this pump is correctly closed. Many pieces of equipment have items that may be considered vital parts. These inspections should be assigned, by name, for someone to do. He/she should be told what to look at and what to look for, as well as how often to do it. Vital parts should be marked as such when practical to do so. It is a good idea to make some plastic-coated vital parts cards for those who have the responsibility to make these inspections. These will serve as a tangible reminder to those who do this work.

SUMMARY

Most farms have a relatively small number of workers, only a few of whom are full time. These people are normally quite familiar with equipment, methods, and other personnel on the place. It would be relatively easy to make the assumption that inspections are not needed, because everybody knows all about it. In this case, familiarity breeds a false sense of security. Planned inspections give system and meaning to hazard recognition and abatement. We do not inspect only when we know there is danger; we inspect when we don't know!

To further keep familiarity from becoming a problem, many proprietors change areas to be scrutinized by the inspectors regularly. What a familiar person may walk right by and not see, one who is less familiar may spot the hazard at once. Heavy industry plants regularly shift inspector personnel. They have been in the hazard abatement business a long time. Farm personnel may learn from them.

Hazard classification, even a simple system, helps establish a priority for correction. Classifications on some hazards may change, depending upon the season of the year. For example, a broken windowpane in a hog brooder pen may be a "C" in summer, but may be an "A" in the winter.

While informal inspections take place daily or weekly, formal ones are normally used prior to busy seasons. Whether formal or informal, all inspections are planned, systematic work designed to identify hazards which could cause losses within the operation. Once identified and classified, hazards should be corrected as soon as it is practical to do so. If hazards cannot be abated right away, people should be warned about their existence. In some cases it may be proper to rope off the area containing the hazard and/or hang a danger sign at the area.

Inspections are a safety tool that has been around for a long time. Proper use of this device can have the effect of reducing losses and saving organizational resources.

FURTHER READING

Lack, R. W., Ed., *Essentials of Safety and Health Management*, CRC Press, Boca Raton, FL, 1996.

CHAPTER 11

Accident Investigation

Why investigate accidents? Everybody knows what happened! These statements may come up after an accident has occurred and someone suggests that it be investigated. The only true reason for accident investigation, as viewed here, is prevention of a recurrence. What is an accident?

An accident is an undesired event that results in physical harm, damage to property, and/or degradation to the environment.

It is unwise to investigate an accident for any reason other than prevention. On some occasions, accident investigations have been used to place blame or to find fault. When used for these purposes, accident reporting is driven underground and the real purpose, that of prevention of similar accidents, is not served. Since prevention of accidents is a key element in the safety effort, the task of investigating them is one that must be accomplished. Accidents that do not result in measurable loss are often called incidents. It may be desirable to investigate an incident that had the potential for major loss, since the causes of incidents and accidents are often the same.

Accidental losses may take several forms. Injury to people and animals, fire, equipment and building damage, or chemical spills are some of the losses that may be experienced in a farm setting. Things learned from accident investigation efforts will assist in preventing these types of losses in the future.

Which accidents should be investigated? Some say all accidents. To investigate all accidents is probably not necessary. Accident investigation is a formal activity. It takes time and involves all persons that were present. If a worker is handling a ladder indoors and breaks a light bulb, should this be investigated? Probably not. A proprietor may wish to keep an accident log of all such events, but investigation is usually reserved for cases that cause serious injury or property damage estimated to be over a given amount. In some cases, a very minor accident may have some ramifications that would cause a proprietor to decide to investigate it. Some proprietors investigate only injury accidents that involve loss of time and/or property damage in excess of $1000.00, but they may investigate others if they feel it would aid in prevention.

WHO SHOULD INVESTIGATE ACCIDENTS?

Often, the proprietor investigates accidents that happen on the farm. He/she may appoint someone else to do so, such as one of the permanent staff who has supervisory responsibilities. Permanent people often are more familiar with the operation than temporary employees and are therefore in a better position to assess the situation. Since any corrective action will normally involve these key people, it is proper that they be involved in the investigation.

ACCIDENT REPORTING

Before a decision to investigate can be made, the accident must be reported. Sometimes people are reluctant to report an accident, perhaps for one or more of the following reasons:

- Fear of disappointing the proprietor or of receiving discipline
- Not wanting to be the center of attention
- Fear of damaging their personal reputation
- Fear of medical treatment
- Not wanting to be involved in a lengthy investigation procedure
- Failure to understand the importance of investigation

To a given worker, an accident may be a rare event. He/she may have never been involved in one. Workers may have heard horror stories from someone who was involved in an accidental loss. It is not hard to understand why a worker may be reluctant to report an accident event. Most do not want to be involved and would like the happening to pass with as little disruption of the work or to their peace of mind as possible.

In order to encourage prompt and accurate accident reporting the proprietor should make it clear to everyone that the only reason accidents are investigated is to prevent recurrence. Few wish to become involved with any part of an accident, but if they can see the need for involvement, the task may be less objectionable. Some proprietors get angry when an accident is reported. This is a mistake. The proprietor should control the situation and thank the person who reported the accident. Such positive behavior on the part of the proprietor will help promote the cooperation of those involved in the investigation. Cooperation of the people involved will help move the investigation and remedial action, if any, quickly to a conclusion.

QUESTIONS ABOUT ACCIDENT INVESTIGATIONS

What is an accident investigation? An accident investigation is an analysis based upon facts which are determined as part of the process. Investigation data may include opinions and statements. Almost all data discovered during the process will be helpful in forming conclusions and deciding upon remedial action. It is easy to jump to conclusions during the investigation process. Often conclusions are wrong or incomplete. Wrong solutions to problems usually only make them worse. Once all the obtainable facts are in and analyzed, cause(s) may be determined.

What is the best time to investigate an accident? In most cases, the best time is as soon as possible. Facts are fresher and memories are sharper. Early investigations may begin before the accident scene changes and evidence goes astray. Investigations may be delayed to care for injured or to stabilize the situation, such as plugging leaks, stopping loss, or keeping the accident effects from getting worse. Never hold up treatment of the injured. Take care of them first.

ACCIDENT INVESTIGATION STEPS

Almost all formal work activities, including accident investigation, have certain definable steps. Steps give system to the process and help ensure that time is used profitably. Accident investigations take time, but this is time well spent. Certain of these steps are those that were developed and refined over the years in many types of industry. If used, the following steps will save time and help to arrive at conclusions in a professional manner.

Step one is to size up the situation. An accident investigation should be conducted at the scene if possible and as soon as practical. Work at the scene aids recall by those who were involved. In reviewing the scene, note what is there as well as what is not there. This step is sometimes called protecting the scene. If the scene contains many elements, it may be desirable to take a few pictures. These may come in handy later, particularly if the scene has to be altered. An accident scene on a public roadway, for example, will have to be cleaned up so as not to interfere with traffic. Photos may come in handy in such cases. An important part of this step is to get to the scene as soon as possible after the accident. By sizing up the situation, the investigator will have a better idea of where to start and to set priorities for interviews.

The second step is interviewing witnesses. The main purpose of interviewing those involved is to determine exactly what happened. Getting the story straight is a major key in a successful investigation. By sizing up the scene and interviewing those involved, the investigator should be able to develop a time line for the accident. It is important to determine the sequence of events. Unless indicators show otherwise, the most knowledgeable person should be interviewed first. If this person has been medically evacuated, his/her interview will have to be done later. Because some of the involved people may be defensive, an adversarial interviewing technique is a good thing to use. This technique is similar to the one that the police and the IRS use. The adversary interview is designed to obtain information under somewhat stressful conditions. The sequence of this type of interview follows:

- Put the interviewee at ease. Another way of saying this is to try to get the stress out of the situation. A good way to begin is to assure the individual that the purpose of the investigation is to prevent a recurrence and ask for his/her help.
- Conduct the interview at the accident scene if possible.

- Interview in private, if possible. If there are several people at the scene call the person aside and discuss the accident. With some people, the accident scene may have such an emotional load to it that the interviewee will have to be taken away from the scene. The investigator will have to use his/her best judgment in this case.
- Ask for the individual's version of the accident. Tell him/her to start at the beginning and go through the story. Do not interrupt nor finish sentences for the person. Wait the person out and get the most understandable story possible.
- Ask questions about his/her version if it isn't clear. Try not to ask why questions. They call for a conclusion and the person may feel trapped. Ask whom, when, where, what, and how. Save the why for later.
- When questions have been resolved, ask the person to tell the story again. When the investigator has no further questions, he/she should repeat the witness's story and ask the witness if he/she agrees with the investigator's version. If the answer is yes, the interview is complete.
- Thank the witness for the help. Tell him/her that the information received will help prevent future accidents.

The investigator should keep good notes of all interviews. It normally does not bother a witness if the investigator is writing while the witness is talking. If the witness seems uneasy about it, show him/her the notes or stand/sit so that the witness may observe them. Allow the witness to read the notes when the interview is complete. If there is disagreement, straighten it out then and there. Ask the witnesses to come forward at a later time if they should remember something further about the accident.

RECORDING ACCIDENT FINDINGS

Normally, completing an accident investigation report is the last phase of the investigation. A form may be used if available. Otherwise, findings can be recorded on a blank piece of paper. Whichever method is used, certain vital data should be included:

- Identify the date, time, and location of the accident. List the persons and equipment/facilities that were involved.
- Fully describe the accident based upon review of the scene and the interviews. Be sure the story is straight.
- Analyze the causes. What were the acts/failures to act and conditions that contributed to the accident. Don't stop there! Endeavor to answer the question, "Why were the acts/failures to act and conditions there?" This is a better and more useable cause analysis than simply looking at unsafe acts and conditions. The latter are perhaps symptoms rather than causes. Complete a good cause analysis and let the chips fall where they may!
- Take remedial action as required to aid in preventing recurrence. If such actions are hard to develop, suspect the cause analysis. A good cause analysis most often suggests remedial action.

A final step is to assign remedial action work to those who are to accomplish it and move on with the work of the farm.

The question is sometimes asked, "What if a person violates a rule and gets injured in an accident? Should he/she be disciplined?" An answer to the question follows. People are normally not disciplined for having an accident. They are normally disciplined for breaking a rule. On the other hand, if a given rule has been violated with impunity, but one day someone who violates it has an accident and is disciplined, a bad effect upon morale may result. It will appear, in this case, that the person was punished for having an accident. The proprietor should apply his/her best judgment to such cases.

ACCIDENT REPORTS

Notification of an accident occurrence is normally done verbally. If the farm carries workers' compensation insurance, a first report of injury will need to be filed. If the proprietor desires, he/she may develop a first report form for use on-site even if workers' compensation insurance is not carried. In most cases, a verbal first report will suffice.

The accident investigation report should be written and kept for a period of time. Sometimes accidents wind up in litigation. If this were to be the case, a written report and any pictures taken would probably come in handy. Legal counsel should be sought in these cases.

ACCIDENT REENACTMENT

Sometime it may become necessary to reenact an accident. This activity can be like playing with dynamite. If the investigation is at an impasse, and if information cannot be obtained in any other way, a reenactment may be required. This process should proceed with extreme care. Do not have a second accident while reenacting the first one. One reads newspaper stories about this happening from time to time. The old admonition, "Do it like porcupines make love — very carefully," should be a primary consideration.

Make sure all involved persons know exactly what is going on when conducting a reenactment. Have the involved worker go through the steps in slow motion, and caution him/her not to make the contact. Do not use a worker who is still upset about the accident in a reenactment. Emotional involvement may cloud his/her judgment.

INVESTIGATING NONINJURY/NONDAMAGE ACCIDENTS (INCIDENTS)

Army Air Corps and other military studies completed in the closing days of World War II concluded that the causes of accidents and incidents were all but identical. What does this mean to us? It means that if incidents (noninjury/nondamage accidents) are investigated, very useful causative data will be discovered. The data may be applied to the workplace as remedial action prior to any accident occurring. Granted, this work takes time, but its results can be very beneficial.

Getting incidents reported can be a problem. Interestingly enough, once workers learn that the proprietor appreciates such reports and that no punitive action will be taken, they usually report incidents. There is little emotional involvement with an incident. Nobody was hurt. Nothing was damaged. Incident investigation is a process that a given proprietor may wish to try for a time and then decide whether or not to continue the activity.

SUMMARY

Unlike safety communications and inspections, accident investigation is an after-the-fact work activity. The accident has happened and a loss has occurred. Accidents make people uncomfortable and the quality of work may be affected for a time following the accident. In order to try to prevent a recurrence, serious accidents should be investigated. It is relatively easy to go through the motions in an investigation and come up with few or incorrect conclusions. Older safety technology held that accidents were primarily the result of unsafe acts and unsafe conditions. Later work has shown that unsafe acts and conditions are often factors leading up to an accident, but they are really symptoms, rather than causes. Unsafe acts and conditions are often called immediate causes, while the reason(s) for the existences of the immediate causes are the basic causes of an accident.

Following is an example of an improper accident investigation: A worker gets a ladder out of the storage area and uses it to climb up and reach a broken light bulb. The ladder has a missing rung. The employee falls from the ladder and hurts himself. The accident investigation reads, in part:

- What was the unsafe condition? The broken ladder.
- What was the unsafe act? Using a broken ladder.
- What are we going to do about it? Get rid of the ladder and tell the worker to be more careful.
- Case closed!

This is an example of a very poor accident investigation. Almost anyone, whether trained in safety or not, would ask some questions of the investigator:

- Why do we have dangerous and unserviceable ladders around here?
- Who is in charge of inspecting tools and equipment for safety? Is it being done?
- Was the worker told to always inspect tools for safety before using them?

These three questions are proper ones in the case of this accident. It can be seen how limiting using only unsafe acts and conditions can be. The cause of the above accident was a management failure — failing to inspect the ladder and taking the broken one out of service, and a failure to train employees to always perform a safety check on tools and equipment he/she is going to use.

Nothing here is intended to indicate that people should not pay attention to unsafe acts and conditions. When these are found, however, it is usually an indication of something amiss in the management system. When investigating an accident, unsafe acts and conditions seem to jump out. Do not stop there. Find out why they were present. By so doing, the investigator will be getting at the real or basic cause of the mishap.

Unfortunately, most proprietors and farmworkers are amateurs when it comes to investigating accidents. Hopefully, accidents are a rare event on a given farm. If one happens, however, the proprietor and his/her people would do well to thoroughly investigate it and use what is learned in an effort to keep a similar accident from happening.

FURTHER READING

Ferry, T., *Safety and Health Management Planning*, Van Nostrand Reinhold, New York, 1990.
Petersen, D., *Safety Management, A Human Approach*, Aloray, Inc. Goshen, New York, 1988.

CHAPTER 12

Measurement of Safety Work

When asked how their safety program is progressing, many people start enu-merating accidents. "We had a broken arm, two sprains from falls, one animal bite and a smashed gate." Accidents are not safety, and as a matter of fact, neither are lack of injuries or damage. Safety, as it has been discussed in this book, is work we do to keep accidents from happening.

Thus, measurement of safety becomes measurement of safety work. If the work of safety, such as inspections, safety briefings, on-the-job training, etc. is properly done, accidents often decrease. Measurement of safety work is not difficult. It is measured in the same manner as production. This chapter discusses some measure-ment devices, sometimes called checklists or standards, that will assist proprietors in tracking safety progress.

Standards for work are used extensively in many industries. The management function of quality control makes use of standards. Production and cost control are other functions that use standards. The use of standards in safety is a relatively new practice, although many organizations have used them for several years. Standards, when used as checklists, keep track of the quantity of safety work. When employed with value factors (VF) they provide a quality assessment as well.

Where do standards come from? Most often, they are made in-house by propri-etors and their people. No magic formula exists for standards construction. They should be designed to check upon progress of defined safety work, such as inspec-tions, investigations, and various types of training activities, to name a few. Usually, the proprietor sits down with some of his/her people and sets down the various aspects of a given piece of safety work. Usually, the more input the proprietor can get the better and more useable the standard will be. The following standards relate to the safety work discussed in earlier chapters, beginning with some of the work listed under Safety Communications in Chapter 9.

A standard for Safety Briefings is defined as short workplace safety meetings, toolbox meetings, or tailgate meetings. A lot of information is available on how to write a standard. Most industrial engineering texts provide instruction in this regard. For purposes here, an informal system will be discussed in which some knowledgeable people get together and develop the standard. Some questions come to mind regarding Safety Briefings:

- How many briefings are to be conducted?
- Who will conduct them?
- Where will they be held?
- Who will attend?
- How lengthy are they going to be?
- What topics will be covered?
- What preparation and presentation steps will be used?
- How will we keep track of this work?

By answering these questions, and writing the answers in statement form, a useable standard will result. To keep track of each step, a checkmark may be used on the line following the statement. If a quality measure is desired, value factors may be used. The following example will use value factors that represent the relative worth of each of the statements. A 100-point scoring system will be used. This is one with which most everyone is familiar.

SAFETY BRIEFING STANDARD

Safety Briefing Standards Sheet

Standards	Score	Value Factor (VF)
1. A minimum of twelve safety briefings will be conducted per year.	10	_____
2. Proprietor will conduct or assign someone to do so.	10	_____
3. Hold briefings in workplace to last no more than 15 minutes.	10	_____
4. Focus on one topic, assigned by the proprietor or by permission.	20	_____
5. All production people will attend.	20	_____
6. Presenter will (a) Prepare (b) Zero-in on main point (c) Relate to people (d) Drive home the main point.	15	_____
7. Cover topic fully using actual materials or training aids as needed.	10	_____
8. Ask for and answer attendee's questions.	5	_____
Total	100	_____

The proprietor or the presenter may score the briefing by awarding all or some of the points attached to the line. By doing so, a quality score may be obtained. The purpose of scoring is to help the presenter improve as the year moves along. In some industries, a foreman will make the presentation, score him/herself, and the superintendent will score the activity as well. The main objective of the standard is to track progress and to assist the presenters in doing a better job. In some organizations, the standards are given to presenters for their own use and information, with no follow-up by a higher-level supervisor. Whatever method serves the needs of the organization is a proper way to do this work.

Could the above standard be used, as is, on a given farm? Maybe; maybe not. The one above may be changed or modified in any way to better meet the needs of the organization. If the sample value factors used above are objectionable for some reason, change them. Standards are simply tools to help an organization do a better job — in this case, in safety.

Following is a sample standard on the On-the-Job-Training work activity. It may be used as is, or it may be modified to better suit a given situation.

ON-THE-JOB TRAINING STANDARD

On-the-Job Training Work Activity Sheet		
Standards	Score	Value Factor (VF)
Preparation		
1. On-the-job training will be given to workers as assigned by the proprietor.	10	_____
2. Trainer will make a plan to include (a) Breaking job into steps (b) Having everything needed available (c) Selecting and arranging site.	20	_____
Presentation of Training		
3. Go through the job steps clearly and completely.	15	_____
4. Show the person the steps as you go through them.	15	_____
5. Test the trainee by having him/her do it.	20	_____
6. Check trainee's proficiency and follow up.	20	_____
Total	100	_____

The following standard is concerned with safety inspections as discussed in Chapter 10. As is the case with the other standards, value factors will be included. These may be used to emphasize a given part of the standard as needed

SAFETY INSPECTION STANDARD

Safety Inspection Standards Sheet		
Standards	Score	Value Factor (VF)
1. A monthly safety inspection will be conducted on an assigned part of the facility as directed by the proprietor.	15	_____
2. Inspector will look *at* and *for* items and conditions as determined.	20	_____
3. Hazards found will be corrected or classified using A, B, C.	10	_____
4. If A class hazards are discovered, personnel in the area will be warned and the proprietor notified immediately.	15	_____
5. Inspector will note exact locations of hazards found.	10	_____
6. Inspector will make a record of the inspection.	10	_____
7. Inspector will recommend abatement procedure as appropriate.	20	_____
Total	100	_____

The last sample standard will relate to Accident Investigation that was discussed in Chapter 11. It is improbable that a proprietor will wish to investigate all accidents, although a log should be kept on all of them, noting the time, date, place, and loss. The log will serve as a working historical document as far as accidents are concerned. It may also be desirable to investigate accidents where no measurable loss took place (incidents), since their causes are usually very similar to loss-producing accidents.

ACCIDENT INVESTIGATION STANDARD

Accident Investigation Sheet

Standards	Score	Value Factor (VF)
1. Accidents involving medical treatment, loss of time, or property damage when estimated cost of repair is $1000 or more will be investigated.	10	
2. Go to the scene of the accident as soon as possible and secure same.	15	_____
3. Interview witness(es) to obtain information on what happened.	15	_____
4. Use adversary interview technique.	15	_____
5. Determine immediate causes (acts and conditions).	15	_____
6. Determine basic causes (why immediate causes existed).	15	_____
7. Recommend actions to keep accident from happening again.	15	_____
Total	100	_____

Results of accident investigations should be discussed with all workers. They will need to know what happened and how they can help to prevent similar accidents in the future. Accident investigations should be kept on file for whatever use they may have in the future. It may be advisable at a certain time each year, such as before the busy planting season, to review the previous year's accidents and prevention measures with the workforce. Since the purpose of accident investigation is prevention, this action will alert workers to their safety responsibilities.

GOAL-SETTING IN SAFETY WORK

Since the use of safety standards with their value factors scores give a numerical rating to certain safety work, it is relatively easy to set goals for safety. Even though the value factors (VF) add to 100 for each standard, the goal for the year on each may be something short of 100. Many organizations simply use the standards as checklists for the first year to allow people to get used to the concept. On subsequent years, value factors may be used and a numerical goal set for each facet of safety work represented by a standard.

For example, the numerical goal for the first year for inspections may be 85. The following year it may be 90, and so on. The same may be true for other standards. At the end of a measured base period of time, usually a year, the results may be put together to form a total safety score for that particular year. Please see the example below.

Safety Ratings — 1999			
Safety Work	**Goal (VF)**	**Score**	**Variance**
1. Safety inspections	85	88	+3
2. Accident investigations	85	90	+5
3. On-the-job training	85	80	−5
4. Safety briefings	85	85	0
1999 Mean safety score		85.75	

This is a simple example of how safety work standards may be used to measure results of the safety effort over one-year time frame. If someone asks how the safety effort is doing, the answer might well be, "We're running at the 86% level, so we exceeded our goal of 85%." Scores on the various components suggest where to look for problems. In the example above, On-the-job training was the only work that fell short of the goal. By going back and checking the standard sheets on this topic during the year, the low score(s) will make themselves known. Thus, corrective action may be taken to bring the low scores of the standard up to the level of the others.

Safety standards are used to focus attention on the work of safety. Standards serve as guides for those who do the work. Scorekeeping is not only a good management tool, but a motivational device also. In defense of standards, the following question is sometimes asked, "How many people would go to a ball game if they didn't keep score?" Farms keep scores in production — bushels per acre, animals fed out, costs of fuel, and many other numbers are kept. Safety does not lend itself to scorekeeping unless some sort of standard system is used. Of course, accidents can be counted, but these measurements have little meaning. About the only thing that can be done about an accident is to react to it. The action, unfortunately, is always after the fact in the case of accidents. The use of standards makes safety work proactive.

When value factors are not employed, the standards may be used as checklists. Valuable information may be obtained without the use of value factors, such as:

- The number of inspections conducted.
- The numbers of hazards found.
- Causes of accidents.
- The number of people trained for various jobs.
- The number of safety briefings conducted.

Experience has shown that, when standards are used, an organization usually uses value factors also, since these will give a quality score as well as one of quantity. Once the standards system is established, very little time is needed for record keeping. The desktop computer, which most farms have today, will greatly assist in the administration of a measurement system for safety work.

SUMMARY

Because most people are competitive by nature, they tend to pay attention to things that are quantified. Quantification is used extensively in agricultural enterprises in the areas of production, quality, and cost control. It should also be used in safety efforts. The old 100% system that most remember from grade school days is still effective. The military, which pioneered the development of rating systems, uses numbers to transmit scores. Often, the numbers are converted to adjectives when they are reported to the various units, for example, unsatisfactory, excellent, superior, and the like. Army inspection scores are often published for all units to see. High scores often entitle the superior units to fly a pennant from their flagpole. Such tokenism is designed to appeal to the competitive nature of the people who achieved the designation. The pennant is simply a token of reward for hard work. The use of incentives may also apply to farm safety. When the farm's safety rating exceeds the goals set for it, the proprietor may present a piece of clothing or equipment to the workforce with his/her thanks for a good safety year.

Isn't competition necessary for incentives or awards? While farms may not compete with one another the way military units do, they compete against last year's scores. Many industrial organizations enter competitions of some sort, but most agree that the score that means the most is how much they bettered their last year's numbers.

The use of a measurement system in safety, such as standards or guidelines, helps keep safety efforts focused upon the prevention of accidental loss. In addition, measurement tracks progress. Nothing is wrong with having the person who accomplishes some safety work score himself or herself. Unless low scores are used to discipline people, they will usually report honestly. Research has shown this to be true.

The proprietor will ultimately decide if and how safety work is to be measured on a given farm. Experience has shown that when safety work is properly done, accidents usually decrease. Preventing accidents saves money. A measurement system will assist the proprietor in tracking safety progress and will enable him/her to place emphasis on activities that could benefit from improvement.

FURTHER READING

Barenklau, K. E., Effectively Measuring Safety Involves Consequence And Cause, *Occupational Health & Safety*, Waco, TX, March, 1986.

Barenklau, K. E., Developing Standards for Safety Work Activities, *Occupational Hazards*, Cleveland, OH, October, 1989.

CHAPTER 13

Insurance — The Transfer of Risk

Farm proprietors, in addition to their many other duties, have a risk management responsibility. Risk management was briefly discussed in Chapter 1. One of the courses of action open to the risk manager is the transfer of risk. Insurance is a means of transferring risk — in this case from the farm to an insurance organization. Many types of insurance are applicable to a farming operation and some, like crop insurance, are uniquely related. An insurance policy is a document in which an insurance company agrees to indemnify the policyholder and pay losses or claims assessed against the insured up to and including the limits specified in the policy.

Insurance has been around for hundreds of years. Some say that the Chinese first came up with the concept. The story, which gets better with every telling, goes as follows:

> Chinese shipping executives, during the time of Marco Polo or even before, formed a mutual aid society wherein if one of them lost a ship the other members would chip in and reimburse their fellow member for the loss. One day, a wealthy shipper decided he would offer his fellow merchants a plan in which they would pay him a premium each year on each of their ships, and if one was lost, he would pay for it.

This may have been the beginning of the insurance industry. Today, it is possible to purchase insurance coverage of almost every type, including rain insurance for the opening day of the county fair. This form of risk transfer — insurance — is indeed very popular.

"Read your policy!" Policyholders are often given this admonition. Many try it, but probably few succeed. Why are policies so hard to read? Since almost any insurance claim can involve litigation, policies are written by lawyers and designed to be interpreted in courts of law. For this reason most people who do not have a legal background have trouble reading and understanding a policy. A trusted insurance agent is a good source of information on what a policy covers and what it does not. It is often said that an insurance policy covers any kind of losses in the world, except those specified in the exclusions. This may, or may not, be true. Most people rely on the seller for specific information regarding types of insurance and coverage.

WHY PURCHASE INSURANCE?

Insurance is purchased to transfer risk. When the cost of a loss is so great that it might cause the farm to go out of business, insuring for that loss probably makes sense. Of course, certain types of insurance are almost mandatory, such as for automobiles. Many states have financial responsibility laws that require either liability insurance or the posting of funds/securities in an amount deemed appropriate to pay a loss. Without such insurance or other financial arrangement it is impossible to get an automobile inspected or licensed in most states.

Most proprietors insure the farm home and buildings for fire and other types of loss. Policies vary greatly as to what they cover. In times past, farms were usually covered by fire insurance. In later years, package policies became popular wherein other coverage was included, such as theft and liability. If a neighbor or a traveling salesman fell from the porch and was injured, the package policy would probably pay the person's medical bills, if any. If a lawsuit resulted from the fall, the insurance company would ordinarily go to court and defend the insured and pay damages within policy limits. In today's lawsuit-conscious society, package policies are perhaps a good deal for the proprietor. Again, be sure to check coverage. Usually, additional coverage that a proprietor feels is needed is not too costly. Today, almost everything is insurable, for a price. Some of the types of insurance available will be briefly discussed in this chapter.

WORKERS' COMPENSATION

Workers' Compensation (WC) insurance is designed to medically and financially rehabilitate an injured worker. WC, as we know it, was developed in Germany shortly after 1900. In the U.S., federal law mandates WC, but it is administered by the states. Administration, costs, and benefits may vary from state to state. The U.S. WC law was enacted in 1911, although many states had their own WC laws prior to this date. The intent of the law was to ensure that injured workers got compensation for on-the-job injuries. In effect, the law says that if a worker is injured, the employer will pay his/her medical bills and give the person some money to live on while disabled. Prior to the advent of WC, it was relatively common for an employer to deny a claim made by a worker. The employer, to deny claims, could use three defenses from English Common Law. These are

- Fellow servant
- Assumption of risk
- Contributory negligence

WC legislation did not erase these common law tenets; it simply said that they could no longer be used to keep from paying workers' claims. These defenses may have worked as follows.

Let us suppose a worker cut off the tip of a finger while installing a mower sickle. When he went to the employer for treatment and living money, the employer told him that his claim would not be paid because a fellow worker moved the sickle that cut off his finger. "If you want to sue somebody, sue the fellow that moved the sickle." The employer may have used the second English Common Law as a defense, assumption of risk. The organization may have said, "We won't pay the claim; he knows that farming is a dangerous business. He assumed the risk of injury when he applied for work here." Contributory negligence may have been used in this way: "Yes, he lost a finger, but he was goofing off. He contributed to the loss himself; therefore we will not pay."

These Common Law defenses can no longer be used in WC cases. If a worker is injured, the employer pays. WC is an instance of the no-fault concept that, in recent years, has been a topic under consideration in automobile insurance.

MUST FARMS PURCHASE WORKERS' COMPENSATION INSURANCE?

Answers to this question vary from state to state. In general, a farming operation similar to the one envisioned in this book (father, mother, grown son, and up to three seasonal employees) is not required to have WC. Coverage can often be obtained, however, providing the farm meets certain other considerations. If a farm has several full-time employees and generates a sizable payroll it may be required to have WC coverage on its employees. All states have a commission or other body that regulates WC. Specific information as to requirements and coverage may be obtained from them.

WORKERS' COMPENSATION PREMIUM CALCULATIONS

WC premiums are calculated on the basis of so much per $100.00 of payroll. Premiums may also be based, in part, on the loss experience of the insured entity. This means that an insured with sizable losses may pay a higher premium than one with fewer losses. Premium calculation formulas, provided by the state in which the entity is located, usually penalize more for frequency than for severity. Simply stated, an organization with ten $1000 losses may pay more premium than one with a single $10,000 loss. Large losses may place an organization into a debit category wherein premiums may be larger than the normal rate.

In most states WC insurance is purchased from insurance companies. Some states own the WC system and this means that policies have to be purchased from the state. In a couple of instances, WC may be purchased from either the state or private insurance. Rates for WC are set by state formula, so the only advantage in shopping around is for service. It is possible to self-insure for WC. Rules for this process are set by the various states.

SECOND INJURY FUNDS

These funds are a part of the WC system and are designed, in part, to encourage the hiring of partially disabled workers. Suppose a proprietor hired a worker with only one eye; the worker had lost an eye in a previous employment. The proprietor knows that if the worker loses the other eye, he is faced with a permanent total disability case that can be very costly. If the proprietor has WC insurance, and if he/she registered said worker with the WC Second Injury Fund, liability for the second injury (blindness in this case) is limited to perhaps a couple of years of WC payments. Details regarding registration and payment liability vary from state to state. In Minnesota, for example, the above injury would result in the employer's WC having to pay 52 weeks of wage benefits and the first $2000 of medical costs. The fund is maintained by assessments against WC insurance carriers in accordance with state formulas. State rules for second injury funds change from time to time. Second injury funds may only be used if the organization has WC coverage.

WORKERS' COMPENSATION, AN EXCLUSIVE REMEDY

It was the intent of the federal WC law that this insurance coverage provided an exclusive remedy for both the worker and management. The worker would have no-fault coverage and would have medical bills and other payments made promptly. Management would be immune from lawsuits involving an injured employee. Both would be spared legal expense. Today, it is possible for a worker to sue an employer, even though the worker is covered by an employer's WC policy. Courts will hear such cases. In order for the worker to collect, he or she must prove to the satisfaction of a court that the employer was grossly negligent and perhaps intended to harm the worker. Many cases reach the courts, but some are very hard to prove. The law on these cases varies from state to state.

LIABILITY INSURANCE

While WC coverage is standardized and state law sets the premiums, liability coverage may differ and costs may vary greatly. Liability insurance is designed to protect an individual or an entity from losses brought about by injury or damage to others. The liability insurance one carries on an automobile is an example. Personal liability coverage may be purchased as part of a homeowner insurance package. It is important in all insurance to check for coverage. Who or what is covered, and for what contingencies? A trusted insurance agent or seller can be a good source of information in this regard. In the liability area, almost any contingency is insurable. For example, suppose a farm road leads to a fishing lake that is partly on the premises of a given farm. The road is owned and maintained by the farm. Members of the public use the road to go to the lake, sometimes with and sometimes without permission. What if the road washes out during a rainstorm and a car upsets in the wash and hurts some people. Is the farm owner liable? Probably he/she is! The proprietor may wish to pass this risk on to an insurance company before trouble develops. A contingency is probably insurable for a reasonable price. Without this protection, the proprietor may have to install locked gates on the road that would take time and require additional effort on his part whenever the road had to be used. An insurance person should be able to advise in such situations.

CROP INSURANCE

Many types of crop insurance exist in the marketplace. In some areas of the country, hail insurance for crops is popular. As with other types of insurance, the proprietor must weigh his/her needs against the costs associated. What is the risk, and how much could be lost are questions that need to be answered. With this type of insurance, coverage limits may be adjusted and these adjustments will be reflected in the premium charged. As with other types of insurance, the proprietor is transferring risk. Whether or not this coverage is necessary and for what price is a decision proprietors must make.

HOME OWNER INSURANCE

Today, home owner insurance coverage may be tailored for the type and amount of coverage desired. In years past, fire insurance was the coverage most often requested. Later on this type of insurance was expanded to include home furnishings, personal liability, windstorm coverage, and others. The newer policies are often referred to as package policies. Coverage has been available for many years; it became popular in the 1950s. This type of policy has certain limitations. If the home owner has a lot of jewelry, art collections, gun collections, and other expensive possessions the insurance company may require riders to cover such items. Computer equipment may be included in the regular policy, but it is well to check with the company or agent in order to clearly establish what is covered. Riders to cover collections of art and other items are usually not expensive. Always check before making an assumption that such and such is covered. After a loss is a poor time to find out that there is no insurance coverage on selected items.

Personal liability coverage, which can be a part of the home owner's package, is usually a wise choice. This type of insurance comes into play when someone is hurt on the policyholder's property. The type and additional cost of such coverage may be discussed with the insurance provider. Unlike workers' compensation insurance that has fixed rates and coverage, home owner policies may differ. For this reason it is best to discuss any concerns with the provider. The provider can advise policyholders on any overlap that may exist between general liability policies and the liability portion of home owner's coverage. To save premium dollars, all overlaps should be considered and corrected if plausible.

FLOOD INSURANCE

Some time back, the cost of flood insurance was very high. This is probably because people who lived on high ground very rarely purchased it. Those who lived in areas prone to flooding found coverage hard to obtain and very expensive. Today, flood insurance is available at a reasonable cost, particularly for those who live outside floodplains but who may be subject to a flood loss. A policy that provides $100,000 coverage on a home and $25,000 on home contents costs less than $250 per year. Specifications and costs may change from time to time. Costs of flood insurance today are partially underwritten by the government. Cost to those who live inside floodplains are higher, but policies are usually available. Normally, home owner policies do not cover flood damage.

Flood policies are rather standardized. Most insurance companies that sell home owner coverage also handle flood insurance. Bear in mind that flood insurance becomes effective after a 30 day waiting period. Additional information on this type of coverage may be obtained from an organization's insurance provider.

HEALTH INSURANCE

The cost of health care has been rising rapidly for several years. This rise is reflected in larger premiums for this type of insurance, often with a reduction in benefits. Generally, premiums are lower and benefits are higher in group insurance plans. If a given farm unit is looking for a more equitable arrangement for health insurance coverage, group plans that may be available in the area are well worth considering. In some cases, proprietors report dissatisfaction with their health coverage, but cannot find anything available that is better. On the other hand, systematic searches for better and more affordable coverage have often yielded positive results. National and regional farm organizations may have insurance plans that would provide adequate and affordable coverage. If a selection of deductible coverage is available in a given plan, this may provide the proprietor with an opportunity to save some premium dollars, provided that he/she has sufficient cash reserves to pay the deductible portion of medical bills.

OTHER INSURANCE CONSIDERATIONS

Some policyholders find out too late that a particular contingency is not covered. For this reason policyholders should develop a close business relationship with their insurance provider. Ask questions and get answers. It is much easier to accept or decline coverage when the facts are known. Discussions about premiums are best entered into after decisions have been made about coverage. Some insurance companies sell directly to policyholders, while others sell their products through agents. Both systems have advantages. Many proprietors prefer to deal with agents because they feel the agent can get them better coverage choices. Others prefer direct sellers for various reasons. Usually a good practice is to purchase insurance from a reputable carrier that has the financial strength and personnel necessary to fairly settle claims and provide professional service.

Take care to establish correct values on property coverage. A proprietor purchased a home and lot for his son who was a partner on the farm. The total parcel cost $120,000. The proprietor asked his insurance agent to insure the house for that price. The agent reminded the proprietor that he need not insure the dirt. In checking, the proprietor found the replacement cost of the house to be $110,000. The house got insured and the proprietor saved some money. It is unwise to overinsure a piece of property. It costs too much and the likelihood of collecting the amount is doubtful.

Avoid duplicate coverage. With the broader-coverage policies available today it is possible to pay for the same coverage twice. There is no law against paying twice, but collecting twice for the same loss is usually prohibited. If a proprietor deals with a single agent or direct selling company, overlapping coverage is relatively easy to spot. It's harder to determine when several agents or companies are involved.

What about deductibles? Many types of insurance may be purchased with deductible provisions. Deductibles lower premiums. Ordinarily it makes economic sense to carry as high a deductible amount as can be readily afforded. Zero-dollar deductible policies are rare today. Many consider small losses to be maintenance items. A broken storm window or a cracked headlight probably fall into the maintenance category. A policyholder can usually take care of such things for less money than he/she would have to pay an insurance company to do so. In the last analysis, people carry insurance to protect themselves from severe or debilitating losses. The so-called low dollar stuff is usually best handled in-house.

CLAIMS MANAGEMENT

Some feel that once a claim is turned in to the insurance carrier they can put it out of their mind and go about their business. This practice may prove costly. While it is true that the insurance claims people are professionals in their work, any help that the insured can provide them will probably save dollars. When a loss that is covered by insurance occurs, a proprietor should gather all the information he/she can with regard to it. This information should be passed on to the insurance carrier. Proper aid should be provided for the injured. Any kindness and/or concern shown will often help keep an adversarial relationship from developing. If off-the-farm (third party) people are involved, names, addresses, their insurance carrier, witnesses, and other information that may have bearing upon the case should be noted. These people will probably want the same information from the proprietor. Insurance (claim) limits should not be specified.

If it is appropriate to take some photographs, this should be done. If the farm uses an attorney, depending upon the situation the proprietor may wish to notify that person. It is important for the proprietor or his representative to stay on top of the claim until proper disposition has been made. If information comes up later on during the processing of a claim that may be helpful, notify the carrier. The insurance company will settle the claim, but assistance provided by the insured can save dollars.

SUMMARY

Insurance provides a means to transfer risk that is often beneficial to farming operations. With today's legal climate, liability coverage is almost a must. Property coverage is also an important factor. Without insurance coverage, a fire, for example, may well put a farm into receivership.

Many feel that insurance is too costly and they are willing to take their chances. Others say that they can ill afford to risk years of work and the building of an operation, only to lose it as a result of an accidental loss. Still others feel that even though it is hard to pay premiums, it is worth the price for the peace of mind that comes from knowing that what they have worked hard for is, at least in part, protected. Many would agree that lots of risk in a farming operation is uninsurable, such as dry seasons and unanticipated drops in farm prices.

Few would doubt that careful planning of insurance coverage and how best to achieve it is needed. Transfer of risk — insurance — is one of the tools used by those who are charged with the management of risk. Many proprietors are not familiar with insurance terminology, coverage, or payment options. For this reason they may put off utilizing this tool. It is advisable to develop a close business relationship with an insurance provider, whether it is an agent or a direct selling company. The contact person serves as staff to the proprietor in much the same way as the physician or attorney does. The proprietor is free to heed, modify, or disregard staff advice. The advice, however, should be sought when needed.

FURTHER READING

Petersen, D., *Techniques of Safety Management*, McGraw-Hill, New York, 1971.
Wolff, K. M., *Understanding Worker's Compensation*, Government Institutes, Inc., Rockville, MD, 1995.

CHAPTER 14

Off-The-Job Safety

It makes little difference to the production effort whether a valued worker or a vital piece of equipment was lost on or off the job. The person or equipment is not available for use, and the loss is sure to be felt by the organization. When contemplating accident prevention efforts, the proprietor would do well to consider safety off the job as well as on the job. It appears that people are more likely to have accidents when they are away from the workplace.

The Bureau of Labor Statistics shows that the ratio of being killed off the job as opposed to on the job is approximately 4:1. As to disabling injuries, the ratio approximates 3:2; 9:7 is the ratio for nondisabling injuries, while a ratio of 7:5 applies to workdays lost. Note that in every case, the numbers are greater on the off-the-job side. These national injury statistics would not necessarily apply to farm populations, but the point made is unmistakable: more time and effort could well be spent in the reduction of off-the-job accidents.

In-depth studies of off-the-job accidents are in short supply. Some reasons for the large numbers of off-the-job accidents may be

- People spend more time off the job.
- Usually off-job activities are less structured.
- Off-job activities often involve all family members — young and old.
- Unknown people and factors may influence off-job safety.
- Mishaps may be harder to anticipate.
- Unfamiliar environmental factors may affect the off the job site.

The above items may suggest some opportunities for prevention of off-job accidents. Short safety briefings are a tool to increase awareness of hazards which might be expected for a given off-job activity. These briefings are discussed in Chapter 9. A few minutes spent discussing an off-site activity prior to beginning it is time well spent. Some families do this, and it would serve accident prevention well if more did it. A picnic at the lake, for example, is supposed to be fun. However, hazards are involved in this adventure that should be discussed prior to the arrival at the picnic site. Answers to the following questions may suggest topics for safety briefings: Who else will be there? Will there be boating or swimming? Did we remember to pack safety equipment? Where are the closest emergency facilities?

Traffic accidents usually account for more than half of the off-job fatalities. Because of road and equipment conditions, behavior of other drivers, and traffic congestion, control of the situation is harder to achieve. A few minutes spent in route selection and setting of departure times can aid in avoiding some traffic problems. Even the best drivers may have problems because they have little control of the environment while under way. Professional stunt drivers often report more accidents on the way to shows than while they are performing.

SAFETY CONSIDERATIONS IN THE HOME

It has been said that the average home can be a minefield as far as accident potentials are concerned. Most homes contain chemicals and compounds that are every bit as hazardous as those found in industrial workplaces. While it is true that the quantities of these substances in the home are relatively small, a hazard to health and well-being still exists. Items such as insecticides, drain cleaners, and bleaches are often stored in the home. These compounds may cause acute physical problems if ingested or splashed into the eyes. Small children may be especially vulnerable. Parents can attest that babies grow into active toddlers in what seems like a remarkably short period of time. They learn very early on how to open cabinet doors and how latches work. If this weren't enough, small children like to taste things. That is one of the ways in which they learn. They have been known to eat or drink hazardous materials that often result in very serious problems. All hazardous materials should be kept where small children cannot gain access to them. Adult household personnel should be cautioned to store hazardous materials when they have finished using them.

Slips and falls are often the cause of injuries in the home. Overreaching on ladders and using makeshift platforms to stand upon often lead to injury. It has been noted that people will sometimes take chances in the home that they would not do on the job. Work around the home requires thought and planning, just as does work in other areas. Safety reminders, discussed in Chapter 9, work well in the home as well as in the workplace. Remind family members about the safety aspects of the chores they are doing. Some 40% of disabling injuries suffered off the job occur in the home.

Electrical safety should be a primary safety consideration. The use of temporary wiring can cause circuit overloads and electrical shocks. Some farm homes have been used for generations and may have marginal or insufficient electrical capacity for today's increased use of electrically powered equipment. Some electrical outlets may not have provision for grounding and may not accept three-prong plugs. Polarities may be reversed, which can cause damage to some electronic appliances. Testing devices for grounding and polarity may be purchased from hardware or department stores. These inexpensive and easy-to-use devices will test for these conditions. A grounded plug on an appliance is useless unless the circuit itself has a provision for grounding. An electrician should be consulted if problems are found.

VACATION SAFETY

It has been estimated that some 60% of off-the-job fatalities occur in traffic accidents, as do some 26% of disabling injuries. Many people travel long distances while on vacation, so highway traffic exposures are present. Prior planning is one of the keys to safe vacationing. Many vacationers try to do and see too much for the time available. Others prefer to drift rather aimlessly, following no schedule whatsoever. Planning helps in either case. Planning helps people to anticipate problems and have courses of action for dealing with them.

Be sure to carry health insurance identification along on vacation. Without such information it is becoming much harder to obtain medical services without paying a considerable sum of money up front. If members of the group have chronic illnesses that may require treatment while on vacation, be sure to jot down the names and phone numbers of the physician(s) who regularly treat the person. The physician visited on vacation may wish to contact the home doctor.

If driving long distances, take rest stops to get out of the vehicle and exercise a bit. Change drivers from time to time if appropriate. Use the rest stops to consult maps or directories. This is hard to do while driving and may be very unsafe. It is usually a good idea to carry a telephone while on the road. Many people, older ones in particular, travel with a portable phone to use in case of emergency or for convenience. Today, police in many states have a short number to dial in order to obtain their assistance. The numbers are often posted on signs along the highway and are usually listed on official state road maps available at many rest stops. If a phone is available in the car, contacting police or other help is quite easy to do.

Drinking water from home is often carried on trips by those who have problems with out of town water. This is a small consideration that could make travel on vacation a lot more pleasant. Small children travel better if they have games or books along to occupy them, at least part of the time.

It is a good idea to carry a few tools in the car so that minor repairs and adjustments may be completed without undue inconvenience. Adjustments for such things as a loose battery terminal clamp or an engine belt can usually be quickly taken care of if tools are available. Make sure the spare tire is in good condition and properly inflated before starting on a trip. Be sure that tire-changing equipment is carried in the auto. A set of jumper cables is a good addition to the auto tool kit, as are extra fuse bodies for autos that require them. Always carry a serviceable flashlight with fresh batteries.

A first aid kit is a good item to carry in the car, whether on vacation or not. Treating minor cuts and scrapes greatly reduces the chance of infection. Sanitary napkins make good compresses for the control of bleeding. Some of these may be conveniently stored in a zip-type plastic bag with the first aid kit. Although anything placed upon a cut should be as clean as possible, when someone is bleeding severely, complete sanitation is perhaps a secondary concern. To quote a country medical person who is now deceased, "Infection we can treat; death from loss of blood we cannot do anything about."

FIRST AID TRAINING

Having someone around who has had first aid training is as important off the job as on. Every farm family and work group should have at least a couple of people who have had training in first aid. Many lives have been saved because of prompt and correct first aid measures. Farm settings are often located some distance from professional medical help. This underscores the need for trained first aid personnel. Local Red Cross chapters, fire stations, and medical facilities can usually provide information as to where and when first aid training will be given in the local area. The local mutual aid group, if one is available, may have a certified first aid instructor either as a member or on call. People may also be trained in cardiopulmonary resuscitation (CPR). Those who successfully complete first aid or CPR training receive a qualification card that must be renewed by refresher training from time to time.

Personnel trained in first aid may render assistance to any injured person within the guidelines provided by the training they have received. Some worry that they may become involved in litigation if the person they aid should die or become disabled. This could occur, but most states have Good Samaritan laws designed to protect people from legal problems so long as they perform only the procedures for which their training qualifies them. The story is told of a coach who was successfully sued because he performed a tracheotomy on a student who was choking. According to medical authorities, the coach probably saved the student's life. While the Heimlich maneuver is an approved first aid measure for choking, a tracheotomy is not. The latter is a medical procedure. It is doubtful that a Good Samaritan law would have done the coach any good. While it is true that first aid people do get sued, very few, if any, have been successfully prosecuted so long as they were doing recognized first aid procedures.

THE ORILLIA STUDY

There is apparently desirable carryover from first aid training to loss reduction experience in the workplace and off the job. During a three-year period, one quarter of the population of Orillia, Ontario was taught basic first aid as a part of an 8-hour safety-training course. This pilot project was underwritten by the Ontario Workmen's Compensation Board and taught by the St. John Ambulance Association. Of the many conclusions drawn, a noteworthy finding was that the first-aid-trained employees experienced 30% fewer off-the-job injuries than anticipated, and nontrained employees had some 20% more injuries than forecast. From this study, it appears that there is a positive carryover effect from first aid training that extends to off-the-job activities. This carryover effect is another good reason to have as many family members trained in first aid as is practical.

VISITOR SAFETY

Sometimes, people visiting farm sites are not acquainted with some of the hazards that may be found there. Visitors should be cautioned that animals sometimes react differently when strangers are in or near their enclosures. Visiting children may climb upon machinery and enter buildings as a part of their natural curiosity. Those who are not familiar with a farming operation can get themselves into compromising situations.

Tractors should have ground-engaging equipment lowered to the ground after parking. This would preclude someone from accidentally lowering a heavy piece of farm equipment that could trap or injure a person. It is a good idea to caution children not to play on farm equipment because they may fall off or injure themselves in other ways. Some children like to climb up raised grain elevators to reach a bin. They should be cautioned about such behavior.

Children who grew up on farms are usually quite familiar with heavy mechanical equipment. Many learned to operate such equipment at an early age. Their non-farm counterparts will not have the same familiarity with machinery that the farm children do, and for this reason they may be more vulnerable to injury.

EMERGENCY DRILLS

Fire on a farm may have several consequences. It is important that everyone who lives at a given site knows how to report a fire and how to operate fire-extinguishing equipment. Animal safety during a fire situation is another subject about which farm people need to be informed. It is a good idea to brief farm personnel from time to time about what to do in case of emergency. Duties may be parceled out to various people and all should be cross-trained in case someone is absent during an emergency.

USING COMMUNITY SAFETY RESOURCES

Many organizations such as the PTA, fire departments, service clubs, farm organizations, and police units often have resources that may be used to improve off-the-job safety. In most cases, these organizations are happy to provide a list of things they have or can do to assist in safety efforts. If the neighborhood has a mutual aid group, this organization may be used to schedule community events in which all members can participate. Some of the activities that community organizations conduct are

- First aid courses
- Contests of various types that involve safety
- CPR courses
- Fire safety demonstrations and presentations
- Safety banquets
- Neighborhood watch programs
- Lifesaving training and certification
- Watercraft safety
- Gun and hunting safety courses
- Provision of speakers for neighborhood events

The above list is not all-inclusive. Depending upon the area of the country, courses in snowmobile safety, cold weather survival, scuba diving safety, and many other things may be taught by one community organization or another. Community activities can be more interesting and informative than those conducted on-site because of the opportunity to share experiences and to learn from one another.

SPREADING THE SAFETY WORD

Newspapers, particularly local ones, will usually publish news of community safety activities and projects that are shared with them. Successes in safety should be shared so that others may benefit. When members of the community conduct a safety activity, invite a member of the local press to attend. Often, newspeople are not invited because people think they won't come or put anything in the paper or on the air. There is no guarantee that they will. In this regard, one thing is certain; if they don't know about it they can't publicize it!

APPLICATION OF ON-THE-JOB SAFETY WORK

Many of the safety work activities used on the job apply to off-the-job situations as well. All of the safety communications work is applicable, as are inspections and accident investigations. The latter two activities are discussed in Chapters 10 and 11. A log of off-the-job accidents should be kept and these accidents investigated. Since off-the-job accidents may involve third parties, good investigational data are important.

Inspections may be used as a part of pre-trip preparation. Inspections should cover emergency equipment as well as mechanical and behavioral aspects.

Planning is an important part of on-the-job safety and it has a place in off-job activities as well. Planning helps everyone anticipate opportunities as well as problems. Minutes spent planning will often save hours of stress. Leisure time should not be planned to death, but some planning will assist in allowing off-the-job activities to proceed more smoothly.

SUMMARY

The proprietor who is the accident control leader in production work also has the role of leader in off-the-job safety. Unlike factory work, on the farm the work groups and leisure groups are very similar. The housewife, who often performs farm-type work, is a key person in home safety. All have a role to play in the protection of people as well as keeping farm animals, crops, and equipment out of harm's way.

Losses are keenly felt whether related to on- or off-the-job activities. Accidents have a price tag on them. They can be very costly. Off-the-job accidents may have a distinct downgrading effect on other members of the farm work group. Injuries usually have more of an emotional impact than do damage-type mishaps. When a family member is sick or injured others may have problems keeping their thoughts on their work and may therefore become a target for accidents themselves.

Safety considerations for off-the-job activities are as important as those relating to day-to-day work. All need to be reminded from time to time, that in farm safety, people have to assist one another in work and in leisure. Thoughtful preparation is a key to safely enjoying work as well as play.

Sometimes safety on the job and off-the-job is viewed as two different things. This is not the case. Loss control efforts apply equally to both.

FURTHER READING

Gausch, J., Loss Prevention and Proper Motivation, *Financial Executive*, January, 1970.
NSC, Challenge to Management of Off-The-Job Accidents, Pamphlet Stock Number 123.04-601, National Safety Council, Chicago, IL, 1974.

Mutual Aid Group Charter

Following is an outline from which a mutual aid group charter may be constructed. Mutual aid groups are used to enhance safety efforts of the membership. They may also be used to assist with production efforts. Safety and production successes may be shared with members, thereby saving time and money for those who belong. For example: there may be a need for certain testing and monitoring equipment, the cost of which may be prohibitive for a single farm unit. Through the mutual aid group, costs may be shared and several farm units will be able to use the equipment. In addition, expertise of many types may be shared among members, thus providing educationally sound outcomes to the old adage, "There aren't any of us as smart as all of us."

Name of Group

Name of the group shall be *Clear Lake Mutual Aid Group.*

Membership

Membership is open to all farm units within Clear Lake Township. Out of township farm units may be admitted to membership by majority vote of the existing members. Each farm group has one vote.

Group Officers

Officers, elected by the membership, shall include: chairman, vice chairman, secretary, treasurer, program chairman.

Other officers/coordinators may be added to the elected list as needs require. For example, it may be desirable to have an instrument coordinator who would function as caretaker and loan officer for group equipment.

Meeting Place and Frequency of Meetings

Monthly meetings will be held at the Clear Lake Isaac Walton Lodge at 7:30 p.m. on the first Tuesday of each month.

Dues

Most mutual aid groups have a dues structure in place that is used as a petty cash fund in some organizations. An example would be the purchase of a battery for a piece of test equipment. Other groups may collect substantial dues that may pay for items of equipment, thus eliminating the need for special assessments of the membership. Other groups charge themselves fees for the use of equipment. Funds from this activity may be used to purchase new equipment and maintain that which is in the inventory.

Meeting Activities

Educational activities account for much meeting time. Because of its nature, a mutual aid group is usually able to obtain the services of people and organizations that would be practically unavailable to a single farm unit. Most governmental agencies will supply speakers and discussion leaders at no cost to the group. Agricultural experts from colleges and universities are usually available at a modest cost. Groups usually offer to pay mileage to speakers prior to discussing fees. Agricultural county agents are a prime source of information for groups and are often able to supply names and areas of expertise of others that may be of help in areas of concern. Services of a county agent, for the most part, are free of charge.

Equipment dealers and suppliers will often make themselves available to groups, particularly to those who publicize their meetings and activities.

Publicity

News items relating to mutual aid group activities should be shared with local newspapers, radio, and television stations. Such gives the group visibility within the community. Groups with good visibility usually have little trouble getting speakers for meetings. Successful ventures undertaken by the group should be publicized. The sharing of safety successes is a long-standing practice within industrial trade groups.

As the mutual aid group becomes better known, it is not uncommon for it to be asked to supply a speaker for local, community, or state events. When safety matters are shared a win/win situation is created. Attendees learn from presenters; presenters learn from attendees. Often, local chambers of commerce may ask for presentations from mutual aid groups. These interactions become the basis for cooperation between the various organizations that exist within a given community.

GETTING STARTED

Once off and running, groups usually have little trouble in sustaining their safety efforts. Startup requires the active effort of two or more farm groups who get together and set forth desires and ideas regarding the proposed group. When a couple of farm units see the need for such an organization, a social get-together is often used to get the word out to the other farm units within the township. If reactions are favorable, the mutual aid group may be formed and a tentative schedule of events may be established. E-mail questionnaires and suggestion boxes may be used to amass ideas for the group.

SAFETY IS FOR SHARING

When useful safety information is shared, both provider and recipient benefit. To illustrate this point, if a person has $5.00 and gives it to someone else, the giver no longer has it. There may be a benefit to the receiver. On the other hand, if a piece of safety information is shared, both giver and receiver have it.

Until the past couple of decades, safety information was gladly shared, even among the most competitive organizations. This is not the case any more. Managers have learned that good safety efforts help an organization's bottom line, and for this reason some organizations are reluctant to share safety information with competitors. Although some farm units compete for business, most do not compete the same way that manufacturing organizations do. For this reason, farm unit proprietors are not reluctant about sharing safety information.

Sample Minutes: Mutual Aid Group Meeting

Names of people and products contained in this sample set of meeting minutes are fictitious. This sample may serve as a model for keeping a record of meetings held and topics discussed. Record keeping is a necessity for meetings of this sort and will save time and effort as meetings are held.

CLEAR LAKE MUTUAL AID GROUP
MINUTES OF MEETING HELD AUGUST 1, 2000

The meeting was called to order by Chairman P. Knoles at 7:30 p.m. The minutes of the July meeting, kept by Joseph Johnston, secretary, were presented and approved as read.

Old Business

Ben Wiese announced that he had obtained an Air Force surplus battery-charging unit for a cost of shipping and handling only, $26.00! This machine has the capability of charging almost all the types of batteries we use in our monitoring and measuring instruments. Proper charging of these batteries extends their life some sixfold. Although the machine is an older one, it is in excellent condition and appears to have had little use. Last year we spent over $300.00 to replace batteries that may well have been saved if we had had this machine. In July, the membership asked Ben to obtain one of these surplus units for us if they were still available. Our costs were considerably less than the $250.00 allocated for this purpose. Thanks, Ben.

New Business

Alva Harrigan, county agent, gave a presentation on rat control. Rat infestation, as reported by the membership, has been a problem over the past two years. Ratmorte, a new chemical bait for rats, has been developed by Argone Chemical. This bait requires mixing and is served as small cakes. This preparation has worked well on the common brown rat (*ratus ratus*) and the larger gray Norway rat (*ratus norwicus*). Brown Farm Supply now carries this bait in its pesticide inventory.

Mr. Harrigan asked if we would again be willing to participate in the Farm Safety Fair that will be held at Clear Lake schools in the spring of 2001. Chairman Knoles said that the membership would discuss this and let Mr. Harrigan know prior to the notification date this winter. Our participation in last year's fair included a demonstration of various test equipment we use. Ben Wiese and Sam Haroldson were our representatives to the fair last year.

Rick and Sally Koontz asked for progress, if any, on their idea for a tractor/trailer rodeo wherein participants from our own Clear Lake Group and surrounding townships would be invited to send participants. Those who take part could demonstrate competence in several tractor/trailer maneuvers. The Koontz's attended one of these rodeos which was held out of state last fall. Chairman Knoles reported that his committee had not held a final meeting on this topic as of yet, but it looks to be a good way of promoting tractor/trailer safety. This event would be a first for our area of the state. These events are normally held in the fall, after corn harvest. Chairman Knoles promised a recommendation at the next meeting. Mr. Ben Kapshaw, director of the Future Farmers of America chapter at the local high school, has indicated that his class would be interested in assisting with the rodeo project, if it comes to pass. Barnhart Implement Co. has stated that it would supply trophies for the rodeo, if appropriate.

Secretarial Items

Dr. Tom Olson, Professor of Farm Management, State A & M, has asked us to supply a speaker or two for his graduate seminar on Agricultural Safety. This seminar is held every Thursday evening throughout the semester. A & M has offered to pay mileage and an evening meal for those of our group who may wish to participate. Prof. Olson has agreed to allow us to pick the date (with one week's notice). Lance Regan and Terry Kapshaw indicated an interest in working with Prof. Olson on this. A vote in this regard was favorable.

Mr. Darrow of the Ikes house committee complimented us on the fine care we take of their building when using it. Thanks to all participants.

Adjournment

A motion to adjourn was made and seconded. Chairman Knoles adjourned the meeting at 8:15 and refreshments were then served.

Respectfully submitted:

Don Van Liere, Secretary

APPENDIX **C**

Sample Safety Communication Piece

Following is a sample safety communication. This one involves a brief summary of a happening from which all may learn. Such devices are used to summarize accidents and/or incidents from which a lesson may be learned. When successes or lack of same are written up, many may learn from the experiences of a few. It is well to remember that valuable lessons may be learned from successes, as well as failures. Following is an incident report involving a near miss or close call. No injury or damage occurred so this write-up is an incident report. Had injury or damage occurred, this would have been an accident report.

I LEARNED ABOUT SAFETY FROM THAT!

A Safety Communication from Clear Lake Mutual Aid Group

Incident: entrapment due to bracing collapse.

Location: Buffalo Township, State Highway 19, three miles south of the township line.

Summary: John and Tim Perry, members of our group, came upon the above incident at about 3:30 p.m. last Thursday. A worker was changing a universal joint beneath an ensilage machine in a field that adjoined highway 19. As they approached, they saw a worker under a machine who was waving frantically at them. They stopped and went over to the scene. A worker was trapped under the machine; part of it was resting on his heel. It had been raining and the ground was quite wet. The worker had the machine blocked up with 2 × 4s and skid planks. While trying to loosen a bearing anchor, the skid plank moved and the machine came down, pinning the man's heel to the ground. He felt if he tried any harder to free himself, the whole machine may have fallen on him. He had lain there about 30 minutes before John and Tim came by. They helped to stabilize the machine and freed the man from his entrapment. Everyone involved helped to prevent what may have been a serious accident.

Lesson Learned: It is necessary to take extra care and attention if blocking up machinery when the ground is wet. It is desirable, but not always possible, to have two people performing a job such as this.

Index